# But first . . .

*'Publishing is a peculiarly peculiar business, particularly self-publishing, and it's unlikely that this book will ever be available via normal bookshops, even as a special order.'*

That's what I wrote on this page of the first edition of *Scenes*, and at the time it was perfectly true. All I had to show for a £6,000 outlay was a stack of breadbin-sized boxes helpfully propping up the bedroom wall, each box containing 23 copies (2,000 in all. Work out how many boxes for yourself. I never could), and an eight-foot stack of cardboard wrappers in the Packing Shed. I rapidly promoted myself to Marketing Director, Sales Manager, Advertising Executive, Gaffer i/c Accounts, Mail Orders, and Packaging … and He With Whom the Buck Stops, Sniggers and Runs off Again.

I eventually simplified my job description to 'MD, CEO, Janitor'. Brevity – wit, they say.

Over the next two years I somehow sold all 2,000 copies; partly to local shops, partly via wholesalers, and partly via mail orders from articles I wrote buckshee for magazines. And we sold a dozen or so, via PayPal and the www, to 15 countries.

To my amazement, we eventually covered all our costs and began to make a profit (although if you balance time against earnings I was working for under £2.50 an hour. Ah! The glamour of publishing …). But people *did* seem to like the book, so I was well satisfied. We even took the calculated risk of going for a reprint of another 2,000 copies (this time in boxes of 25, mercifully).

Then four things happened: we ran out of markets and couldn't think of where to find any new ones; Scott Pack of Waterstones said 'I like it!'; Scott put me onto an agent, Stan; and Fiona MacIntyre, the boss of Ebury Publishing, emailed me to say 'I love this book!'. One thing led to another, and to my astonishment, *Scenes* has found its new market … you, dear reader! (Or 'dear browser, possibly' … Oh, come on, do make your mind up. You haven't got all day … things to do …..)

I do hope you enjoy *Scenes*.

All best wishes
    Chas Griffin (sadly demoted to mere 'writer').

(Titles originally considered by Third Leaf Books)

Seventy-two ways of wasting money quickly
A day-tripper's guide to the holy roman empire
'Dull, Wet, and Windy': a back-bencher reminisces
Three more ways of wasting money quickly
Your dog as philosopher

*Our Editorial Team*

# Scenes From a Smallholding

## CHAS GRIFFIN

EBURY
PRESS

This edition published by Ebury Press in Great Britain in 2005

First published by Third Leaf Books, Saron, Llandysul, SA44 5HB, Wales

10 9 8 7 6 5 4 3

Text and cartoons © Chas Griffin 2000

Random House, 20 Vauxhall Bridge Road, London SW1V 2SA

Random House Australia (Pty) Limited
20 Alfred Street, Milsons Point, Sydney, New South Wales 2061, Australia

Random House New Zealand Limited
18 Poland Road, Glenfield, Auckland 10, New Zealand

Random House South Africa (Pty) Limited
Endulini, 5A Jubilee Road, Parktown 2193, South Africa

The Random House Group Limited Reg. No. 954009

www.randomhouse.co.uk

www.thirdleafbooks.co.uk

A CIP catalogue record for this book is available from the British Library.

Cover Design by Two Associates
Text design and typesetting by seagulls
Cartoons by Ken Guy
Cover photos © Anne Griffin

ISBN 0091905079

Papers used by Ebury Press are natural, recyclable products
made from wood grown in sustainable forests.

Printed and bound in Great Britain by Cox and Wyman Ltd, Reading, Berskshire

Miss Daisy's wigs by Mr Snippy of Monte Carlo

*Dedication*

*To Mum and Dad, and Anne, and Paddy and Cait,*
*with love and thanks for enabling and sharing this small journey.*

What the national and international press might well have said about the original Third Leaf Books edition of *Scenes from a Smallholding* if the editors (see pii) had been smart enough to send out inspection copies:

..and they're still shamelessly living together...
*News of the World*

**Welsh Agriculture Hurled into Global Jeopardy by Incompetent Hippies...**
*Daily Telegraph*

...we didn't find a single misprnit...
*The Guardian*

**Cor!! All Them Udders!!!...**
*The Sun*

***Far*** **too many footnotes...**
*The Anti-Footnote Pioneer and Bugle*

***Scenes d'un Petit-Tenant*** **est terriblement bon, actuellement.**
*Paris Soir*

**Intestinalundgastrovisceralische gut!...**
*Hamburger Zeitung*

Мы с верблюдом любили каждое слово
*Newsletter for the West Moscow Home for the Loosely Hinged*

# Acknowledgements

I would like to thank a number of people and entities for their assistance in getting this book into your house/tent. First of all, my family, and especially Anne, for her endless patience, support, and cups of coffee, and occasional shoulder to howl on; Ralph and Jessie, and Gordon and Phyl, for their financial and other help; Ken and Doreen for their patient and long-suffering advice; Alun, for keeping our machines rolling way beyond their Die By Date; ditto John bach for our electronics; Ole for getting us PC'd and online; 'John the WWOOFer' Steer, for all his practical help and good humour; John and David Robinson, likewise; Ken Guy, likewise; all the good WWOOFer-folk, who visited us and helped with the work and enriched our lives; Alan Gear, and Judy Steele, of HDRA for their support and encouragement; also Simon Levermore and Tricia; Fran and John at WWOOF; Peter Andrews, of Ecologic Books for his most generous and helpful advice on publishing; Peter Lock and Katie of Design Elements for their support, and work in designing the book; Jonathan Lewis of Gomer Press for his patient advice on paper, format, fonts, and all that complicated techie stuff.

Diolch i Morwena a Beti hefyd.

And to Bruce, Jan, Dave, Trisha, Jeff, Ruth, Peter, Carolyn, Bernard and Jean for their criticism and, for this new edition, Liz Shankland, Suzanne Jones, Carolyn Ekins of www.acountrylife.co.uk, Sheila and Joff of www.flash granny.co.uk, Scott Pack of Waterstones, Mark Stanton (my excellent new agent), Fiona MacIntyre, Claire Kingston, Sarah B, Diana R, Alun Owen and the rest of the Ebury Posse. And Anabel Briggs.

AND, above all, all the friends and customers who have offered such kind and continuing comment, feedback and support.

Also to Daisy and April and Wally and all the other kiney kin; Cheeky and Rocket and Simon, and the rest of the sheepy people; the pigs and poultry, obviously, except for Adolf the gander, and Ahriman, the Cockerel from Hell … although they did keep us on our toes, I guess, and proved that Hate need not be personal.

And the cats, Twinkle-Ears and Pudding; the rabbits Blackie and Fluff; the, er, gerbils, Chalkie and Nobby, of course; the singing blackbird, the various beetles, voles and arthropoda; nematodes, thrips and small squishy critters, visible and invisible; and … well, you get the drift.

Also, our two lovely dogs, Porky and Dylan, who brought so much fun to the farm; and did their bit for fertility, more often than not just where you weren't expecting it.

I would also like to thank the clouds, the trees … the very grass upon which we walk (CUE VIOLINS) … the air we breathe … Father Sun and Sister Moon …

The entire population of Dyfed, as was; large areas of France, Kyrgizstan, and Mauritania; FitzRoy; Viking; Dogger …

(EXIT, PURSUED BY THE READER, AND NOT BEFORE TIME.)

# Contents

The dates below indicate the issue of the HDRA Newsletter, or 'The Organic Way', that the article first appeared in, although why anyone might wish to have this information eludes me for the moment.

| | |
|---|---:|
| Foreword and Forewarning | xi |
| Introduction and Background | 1 |
|     Garlic – autumn 1989 | 10 |
| Now what …? | 14 |
|     Courgettes – winter 1989 | 17 |
| Cheap at half the price … | 24 |
|     Crop news 1989 – spring 1990 | 27 |
| First things first … and second … and … | 31 |
|     Sheep and lambing – summer 1990 | 37 |
| Heating the breeze … | 40 |
|     Crop news 1990 – spring 1991 | 45 |
| You dirty rat … | 49 |
|     Crop news 1991 – winter 1991 | 54 |
| The image of his mum … | 57 |
|     Commercial vs garden growing – spring 1992 | 63 |
| Timeless … | 69 |
|     Organic vs vegetarian? – summer 1992 | 71 |
| My kingdom for a …! | 76 |
|     Water problems – autumn 1992 | 80 |
| Water, water … | 84 |
|     Muck and spreading – winter 1992 | 89 |
| The joy of spikes … | 92 |
|     Orchard – spring 1993 | 97 |
| A fruity bit … | 101 |
|     What a beautiful place – summer 1993 | 104 |
| Radish squash | 107 |
|     Romantic vs death – autumn 1993 | 111 |
| Learning curve … | 113 |
|     Geese – winter 1993 | 117 |

More than we bartered for ...                                    119
    Runner beans – spring 1994                        124
Of woolies and wallies ...                                       128
    Polytunnels – summer 1994                         132
Plastic Pointers                                                 136
    Raised beds – autumn 1994                         141
That's the way to do it ...                                      146
    Brambles – spring 1995                            150
Just hang on in there ...                                        153
    Radishes – summer 1995                            158
Your first million ...                                           163
    Hay – autumn 1995                                 167
Hay and the other one ...                                        171
    Wind – winter 1995                                177
On battering butter                                              182
    The tractor ritual – spring 1996                  187
Forget your Ferrari ...                                          190
    Rabbits – summer 1996                             197
One for the psychos ...                                          204
    Bees – autumn 1996                                211
Stick it ...                                                     213
    Spuds – winter 1996                               221
Let us spray ...                                                 226
Another fruity bit, but pongier ...                              228
    House music – spring 1997                         236
Three centuries on ...                                           239
    Foxes – summer 1997                               248
N n n n n n                                                      251
    Shearing – autumn 1997                            260
Gated                                                            263
Ruminations                                                      267
    We're all different – winter 1997                 271
Swine                                                            276
    The tale of the kale – summer 1998                282
A little thread of logic ...                                     286
... and a little thread more ...                                 292
Nemesis parks her bike, and ...                                  299

# Foreword and Forewarning

As many as two\* people have recently suggested that I should consider publishing a collection of the 'Smallholding Scene' articles which the HDRA (see page xiv) has been good enough to print over the last dozen years or so, firstly in their Newsletter, and latterly in their magazine *The Organic Way*.

'A book?' I thought. 'Proper *book* … instead of just odd scribblings? Hmm … why not? After all, there would be *guaranteed* sales of two\*.'

And then as my wife, Anne, thought it might be a good idea as well, I took a deep breath and got on with it.

This is the result.

It's not a detailed handbook on how to run a smallholding; it's a personal memoir of how we did it our way, written for fun.

It's not completely void of handy hints though. For example, if you want to know how to put up a polytunnel, or the most efficient way of growing lots and lots of runner beans, flick immediately to the articles called, rather raffishly, I think, 'Polytunnels' and 'Runner Beans'.

And if you want to learn a pretty useful Tractor Starting Ritual … well, it's all in there.

It's a book of two halves, inter-twined: the articles themselves, which are obviously themed; and the rest, which is extra material that first of all elaborates a little on the article's content, and then switches into historical mode to explain why we chose smallholding over accountancy or piracy, and how two utter novices went about getting a system up and

---

\* Possibly three, come to think of it.

more or less running, when one of them didn't know hay from straw (names are named).

The inter-twining might cause readers of a tidy persuasion some heartache. Please try not to let it bother you. I promise you will soon get used to an article on 'Cropping in 1989' being followed by something about making a pig's ear of learning to plough in 1984. And don't get too upset with the occasional overlap of information and timelines. The articles were written in retrospect, sometimes years after the event. Some slight repetition has been inevitable from time to time to time.

The aim is that by the end of the book you should have some sort of impressionistic idea of the fun and calamities we had in our first three years down on Dwlalu Farm. I hope, for example, that you will:

- *Thrill and tingle* to us burning Dyfed down …! (nearly)
- *Wonder and delight* at the watching fox …!
- Be really *quite interested* in how many different sorts of 'stickiness' there actually are …! (yes, we're talking bottling honey, here)

Readers who have a literary turn of mind will be reassured to learn that the book *does* have a beginning, a middle and an end: and Nemesis turns up, bang on cue, smacking her palm with a rolled-up copy of *Woman's Weekly* and the traditional cry of 'Right. Which of you cocky barstads is next?' But you'll have to read through to find out how and when she arrives, and who gets walloped, and how.

I've done my best to be accurate in my factual reporting,

but there are bound to be occasional errors and gaffes for which I apologise in advance, so please don't write and tell me, for example, that the Purple-Sequinned Sardonic Warbler I mentioned *en passant* in the article 'Songbird Snares and Recipes: the Pros and Cons' could not possibly be breeding so far north of Cadiz. I don't care. As I say, this is not a text-book; it's a bit of fun-with-information which I hope might amuse a few people for an hour or two.

So here it is ... three years of history somehow inter-larded with nine years of articles and sprinkled with a posy of cartoons by Ken Guy.

One further point: over the years, the quality of the articles has drifted from Worthy to Tabloid, and is currently spiralling down the Dipstick of Quality towards the plain Whimsical and Silly. I insist, however, that the text is always based firmly upon real events. Except where it obviously isn't: I don't really refill my batteries with custard, for example. Well, just the once, maybe.

*Chas Griffin, Newcastle Emlyn, West Wales,*
*for Christmas 2002.* \*

(Incidentally, the Purple-Sequinned Sardonic Warbler happens to be a bit of a goer even if conditions are not absolutely favourable. This I know for a fact. Northerly latitudes clearly hold no terrors for it, and possibly numerous compensations of which science knows little as yet. Our bathroom bears witness to this. More later.)

---

\* And still in glorious West Wales for this spanking new Ebury edition, May 2005

The *Henry Doubleday Research Association* (HDRA) is the largest organic gardening association in Europe, and does sterling work in Africa and beyond, researching into suitable organic techniques. Their total membership (UK and overseas) is some 30,000, a figure that leaps steadily higher with each successive food scandal. Their headquarters are at Ryton Organic Gardens, near Coventry.

Contact: enquiry@hdra.org.uk or visit www.hdra.org.uk

# Introduction and Background
# Big Acorns from Little Roots

Do you remember the third of May 1980? Of course you do. Lovely sunny day. At least it was in the leafier suburbs of Nottingham.

I left the staff room at 5.30, tucked my trousers into my socks and walked up the few steps to where I'd left my lovely Falcon, beaming out a silent prayer that nobody had pinched any more bits off it, when suddenly I was struck by a thought … that my life was a journey. Shattering stuff, eh? But it was, for me. It may sound like a platitude worn absolutely smooth by repetition, but when something hits you, it hits you.

My two-mile ride home through the inter-war avenues of West Bridgford passed in a blur of excitement.

I parked the bike, smiled at the dog, nodded to the chickens and flounced into the kitchen.

'Wife', I said, 'I have been epiphanised. Suddenly, in the yard outside the College, I realised that My Life Is A Journey.'

'Will egg and chips be alright tonight, only I've had a bit of a day with Caity …'

'Yes … a Journey. Just outside the staffroom, it was, while I was …'

'Here. Could you change her while I peel the spuds?'

\* \* \*

And that was roughly how it was. Not quite as dramatic as I've painted it perhaps, but poetic licence is sometimes called for when a life-changing moment occurs. And it was definitely on the steps outside the staffroom.

\* \* \*

In fact, we'd been slowly building towards this moment for years. For the past decade we'd been living out our unremarkable and contented middle-class suburban life: steady as she goes, solid values, local politics (moderate Labour, what else?), CND … you can probably add a dozen other attributes yourself. All our friends seemed to be teachers or probation officers. And we all had 2·4 children.*

Many of us lived in Lady Bay, the only quite beautiful part of Nottingham's south bank dormitory, and we personally lived in the most beautiful and desirable house in the whole district: a lovely three-storey Edwardian villa on a corner of Holme Road. Our neighbours were fine, the district was fine, our income was fine, my job was fine, the College was fine, the students were fine.

We even had an allotment 30 seconds from our big black back gates, on the floodplain of the Trent. A two-minute walk took us and the dog to the river bank; and from there it was a five-minute stroll to Trent Bridge or the City Ground, to watch Forest in their glory days. A quarter of an hour's gentle pedalling took us to Holme Pierrepont watersports centre; and I could cycle into the city centre in under ten minutes.

So what was all this 'epiphanised' nonsense?

Is it possible to be too content? Is cosiness too close to sterility? And is sterility too close to death? And am I being over-pretentious here?

Maybe; but despite having and enjoying all the civilised qualities of life provincial England could offer, there arose a

---

* Not quite true: one neighbour had only 2·3.

faint and indefinable sense that something was missing. No, not quite 'missing'; more sort of 'unrealised as yet'.

In retrospect it is perfectly obvious that the 'problem' and the 'solution' had been developing simultaneously over the last few years. I had begun as a teacher of languages, then moved into English, then General Studies, then English as a Foreign Language, and then into Communication Studies. I seemed to be changing my subject with increasing frequency. This was the symptom of the 'problem': a sort of vague and generalised unease.

Meanwhile Anne (She Who Understands Things) had managed to engage my interest in gardening. It had taken a long time, as I had grown up with the idea that gardening was at best an effete occupation, and at worst an effeminate affectation on a par with interior design and knitting. I had no interest whatsoever. But when Patrick was born Anne managed to somehow persuade me that growing a bit of our own veg would save some money. I was consequently gently elbowed into scattering a packet of carrot seed into the border round the back lawn. 'Sand and ants' is all I remember from this introduction to horticulture. And a dozen or so limp and pallid carrots, somehow ambushed by the ants, I think.

But the thrill of the seed had been sown. Those sprouts; that stem; those leaves; then that root … All came from that tiny dot of a time capsule. Wow. Gradually I got more involved with the prospect of free food.

Anne meanwhile drew my childishly scattered and frivolous attention to the first stirrings of the Green movement. Lots of things now began to click into place for me: the relationship between pesticides and illness; how Big Business was swindling

Third World farmers with false promises of chemical bounty; why 'fresh' veg from the shops didn't keep and why carrots tasted of mothballs; how food and politics and business were inextricably intertwined; and what that oddly intestinal word 'organic' had come to mean. This was the symptom of the 'solution': our life must somehow change its emphasis or direction.

We got one of the new allotments just outside our gate and learned from scratch some of the mysteries of the soil and plant symbiosis. We read books on composting and pH, and cloches and protection, and the use of irrigation, and cabbages and things. And the more we learned, the more and better crops we grew, and the more we came to realise what an astonishing world it really is, that plants should grow at all, and that the organic way was really the only sensible and grown-up way to do it: by enabling Nature, not forcing it. So we joined Henry Doubleday's mob and learned some more.

One day in the autumn of 1979 we sat down to a meal, which contained 21 different homegrown crops; 22 if you count the elderberries in the wine. I can't remember all 21 items now, but they ranged from potatoes and beetroot to parsley and chives. And eggs, of course, from Brenda and Mabel and Pippa and Gladys, the coven of Rhode Island Biddies we had recently corralled under the apple tree in the garden.

Gradually, growing the veg and banging bits of wood together to make a hen slum became more interesting than teaching teenagers the ins and outs of transparently spurious Theories of Communication for exam-passing purposes.

And I discovered that I liked digging; and in fact I enjoyed it rather more than I liked explaining the simple rules of the apostrophe for the fourth time to an amiable gaggle of secretarial girls whose only real interest in life was what they were going to wear on Saturday night.

And I found I loved the challenge of planning the four-plot rotation to maximise the health and volume of next season's crop far more than I relished the endless stream of pointless forms and registers that Management kept thrusting at us.

Gradually the two sets of symptoms were converging to the point when even someone as dense as me would have to recognise what was going on, and make the connection.

And that is why the third of May 1980 was so epiphanical.* At that moment when I realised my life was a journey, I also realised, with a jolt of paradox, that I'd learned everything I could from teaching, and that my Journey was calling me to move on. If She Who Understands Things was agreeable, of course! Time to talk.

* * *

It will come as no surprise to any woman who reads this that Anne had had it all worked out for months, and had been steadily drip-feeding me with hints and suggestions on the hour every hour for several weeks.

So she agreed to talk; and we talked and we talked and gradually realised that several things were coming together to form an extraordinary window of opportunity, through which we could leap into the excitement of the unknown, or which we could slam tight, to carry on until retirement, dotage, and death in our safe and bovinely contended ways.

---

* Or possibly 'epiphanoid'.

Make-your-mind up time was approaching, and, once recognised, would not go away.

The key element, apart from my fading interest in 'education' and growing interest in growing was that Paddy was now nine, and Caity was coming up for two. In a year or two they would be ready for major educational transitions. Thus, any move we might make must coincide with this transitional period if we were to minimise the disruption in their lives.

Also, my parents were quite likely to be taking early retirement from Kent within the next 18 months or so. It would make sense for us to pitch in together somehow, as long as we could agree on such essentials as separate kitchens and living space. Pooled resources would buy better premises.

By now I'd expanded our activities into the bottom few yards of the vicar's garden. Peter and Lindy were happy to see the land well used, and I was happy to experiment with growing garlic in a sheltered spot. It worked well, and Anne and I began to piece a plan together.

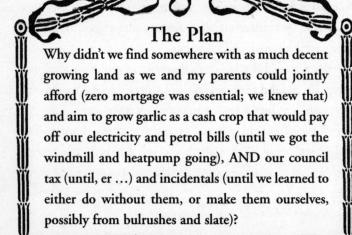

## The Plan

Why didn't we find somewhere with as much decent growing land as we and my parents could jointly afford (zero mortgage was essential; we knew that) and aim to grow garlic as a cash crop that would pay off our electricity and petrol bills (until we got the windmill and heatpump going), AND our council tax (until, er …) and incidentals (until we learned to either do without them, or make them ourselves, possibly from bulrushes and slate)?

So we began to become moderately expert in the cultivation of garlic.

Why garlic? Several reasons: it's compact, stores well, travels well, pays well per unit, can be processed in various value-added ways and, on the non-commercial side, smells terrific and is vampire-resistant. And, from a practical horticultural perspective, it can be planted in the winter, before the spring rush; AND requires a lot of hand-work, thus being suitable for a small labour-intensive set-up. So … there we were.

**The proposition on the table is: that we should look for a smallholding where we can grow our own food, and pay for anything we can't bodge ourselves with cash derived from organically grown garlic. What could be simpler?**

\* \* \*

It took a long time before we got as far as this simple proposition, of course. Was I really that fed up with College? Wouldn't I miss my colleagues? The students? The invigorating five-a-side football after the purgatory of Mechanics III from 4–5pm on a Friday afternoon?

How would Anne feel about not just being able to nip down to the supermarket?

How would the kids feel about having to move from our lovely home; and from school?

Lots of things to talk about and mull.

What about my outside interests at Radio Nottingham and The Scheme folk club?

Would we miss going to the theatre and cinema? Nights out? And friends, of course? Especially our neighbours, Bernard and Val? We talked and talked.

7

The key point for me was probably the realisation that the job I had, and the house, and the neighbourhood, and the general situation of our lives were about as good as it gets in conventional terms; certainly we were not complaining. I couldn't imagine anything better, especially as nobody had ever shown the slightest interest in promoting me (not the Right Stuff, I fear). But ... to stay meant to stay forever, until somebody much younger, and wearing a much snappier jumper gave me a modest whip-round cheque in the Refectory one brilliant June morning some 25 years hence, and we all trooped down to the Wolds Hotel for the customary farewell halves all round, then ... what? 'That was my life, was it?'

## Boring Old Ex-Teacher Dies. Yawns all Round

### No Pictures pages 6, 7, 9

Some people could look back upon having sailed round the world in a jamjar; some had invented wonder-drugs; some took extraordinary photographs; and what would I have done? Churned out more or less pointless information year upon year for a quarter of a century, to young people who had better things to do on the whole, but who had to jump through more or less irrelevant examinatory hoops or die.

And when St Peter sat me down on his knee, and let me stroke his beard, and said, 'Well, little man, and what did we learn in our 75 years on earth, ho ho?', all I would be able to say would be 'Everything there is to know about the apostrophe, and four or possibly five utterly fraudulent Theories of Communication.'

It eventually became clear that there was only one option: to choose Life over Death. But what about the details? Mum and Dad seemed to be agreeable. They had already planned to move north to be nearer the grandchildren. So that was OK. They would move in with us while we planned the final details.

Where, then? We'd read our John Seymour and had a rough idea about the apparent realities of smallholding life, and that the holdings were not as easy to come by as you might hope. We looked in the local papers: none.

We looked further afield in the county: none.

We asked estate agents: they smirked; one sniggered.

Eventually we realised we had a practical choice of either Lincolnshire, Wales or Scotland. Say 'Scotland': say 'midges'; and possibly 'pterodactyls'. Didn't much fancy Scotland. And what about long dark winters?

Wales? What comes to mind about Wales? Hills … sheep … rugby … daffodils … singing; not bad. Oh, and 'good humour'. And a mild maritime climate. A bit far away from friends, but not as far as Scotland, even if we can hack our way out through the bawling hordes of pterodactyls. Wales a possible, then.

Or Lincs? We quickly realised there wasn't really as much choice in Lincs as we'd thought. And anyway 'very flat' came to mind and not much else. One of our friends came from Lincs and was very glad he had.

So … Wales, then. And pretty soon, it would seem.

**We are about to enter the time warp. Hold tight. No harm will come to you …**

# Garlic, Autumn 1989

This is the first Smallholding Scene article, as it appeared in the HDRA Newsletter for Autumn 1989, complete with its original cautionary billing from the editor:

*Chas Griffin is an HDRA member living and working in Wales. In this article, the first of a series, we learn of the harsh realities of life on a smallholding. So, city dwellers planning a retreat to a rural idyll read on …*

We came to Wales to grow and sell organic garlic and live the simple life. Seven years later there is no garlic on the premises and life is immensely complicated.

This is not a problem. Our early ideals have altered and adapted over the years as logic and necessity intervened. The garlic, for example, disappeared in 1985. We had built up our stock to sixty thousand plants and were looking forward to an income from them. Then the summer struck. Endless rain. The weeds choked the plants; it was impossible to hoe or lift the crop, such as it was; the few plants we did salvage rotted in store. The moral we drew from this was 'The Weather Can Be Unreliable'. We started looking for other cash crops.

This *was* a problem. As we only have about an acre of vegetable growing land it is important to grow something that needs a lot of hand-work that big farmers can't afford to pay for, which returns a good price, and which will grow reliably in our rag-bag 'climate'.

Having decided on the crops to grow we then had to decide on how to sell them for the maximum return. We tried everything. Selling at the gate was a dead loss because we're half a mile from the main road and casual trade is nil. The other problem was that customers inevitably called while we were worming the sheep or seeding in the uttermost corner of the field. They would bang on their car horns till one of us arrived, buy their 2lb of carrots and then go away convinced we should have sold them cheaper because we had such a lot in that big bag.

We tried pre-packs, for sale direct and via the Women's Institute (WI) Market, but the fuss and fiddle of weighing, bagging and labelling for no guaranteed return was more trouble than it was worth.

We sold to hotels and restaurants for a while. We got a proper retail price but it needed a ridiculous amount of travelling, and despite their fine words, if chefs could buy in at a penny cheaper, they would. It doesn't matter how fresh or organic a veg is once it's drowned in Bisto. The customers don't care anyway!

We eventually reached an impasse. If we sold at retail price the wasted time was enormous. If we sold to greengrocers we had to accept a 33 per cent price cut. What is more, they wanted regular supplies, and in quantities we couldn't guarantee. Time to get our act in order.

By divine intervention an organic growers' co-operative turned up.

We squealed at the joining price, which was geared to farmers,* but we joined. This gave us the stability we needed. At least we had a steady market, albeit only paying wholesale prices. Now all we needed was a product.

---

* I still don't see how 'Small is Beautiful' can be reconciled with charging a smallholder with one acre the same entry fee as a farmer with forty or a hundred acres. The small man is immediately at a disadvantage, is he not? Or is my logic at fault here?

Obviously it would be silly for us to grow farm-scale mechanised crops like potatoes or swedes. We had to go for something labour intensive. Three years of thinking, discussion and experiment have found us more or less settled on courgettes, radishes, radicchios and Jerusalem artichokes.

Are we making a living? No, not yet. But every year sees a small and steady improvement.

Regrets? No. We've made remarkably few mistakes. There have been plenty of 'learning opportunities' though!

Happy? Yes, when we have the time.

Ambitions? 'To grow and sell organic garlic and live the simple life.'

* * *

Yes … the garlic was a bit of a blow-out after all. Gosh, we tried though. Sixty-thousand plants and five years' development is a lot to lose, especially as we were just about to begin a proper schedule of controlled selling. The idea was to gradually build up the quality and then the quantity of our strains, meanwhile getting rid of the weaker stock as best we could, via various value-added avenues.

We tried all sorts of ways, and they all had their own attractions. We sold in normal bulk to a couple of greengrocers; in lovely gleaming plaited strings to everyone from grocers to souvenir shops; we developed ways of mincing and salting smaller bulbs, and selling them, probably illegally, in old Cow and Gate babyfood jars we brought down with us for this specific purpose; we dried and powdered some, and sold it in little polythene zippable sachets to anybody we met, somehow avoiding the attention of the Drugs Squad; we sold a few trays wholesale, in hurriedly customised and liveried tomato trays.

Certainly the crop was sellable, and of very good quality. Several of the restaurants we supplied commented on how rich and aromatic our bulbs were. Even the powdered stuff was a cut above. It could be done.

We wrote to the French Embassy, who put us in touch with their Agronomy department, who passed us on to some garlic growers. They advised us on strains, and sold us a few kilos to grow on from. One of the growers even drove 150miles from Oxford when he was visiting his nephew, to see how we were getting on. (I don't think he was terribly impressed, actually; but then his farm was on the fragrant plains of Provence, not halfway up a Welsh hillside, and anyway I didn't much care for his shirt.)

I even had plans to make and sell Garlic-Flavoured Cornettos on the beaches of Cardigan Bay, out of the old ice-cream tricycle we'd brought with us in 'stripped-down condition' ready for renovation.*

One restaurant we approached was le Gavroche. The boss seemed to be impressed with the sample we enclosed and suggested we drop by to discuss terms 'next time we were in Town'. 'Next time'? We never went to London, and it would cost a fortnight's income to make a special trip.

And then it rained, anyway. So that was that.

---

* Which somehow, like The Mechanic's Car or the Builder's House, never got properly finished. It remains in the Woodshed to this day: a little piece here; a larger piece there … most of it hopelessly lost. I don't think Anne is all that distraught, actually.

# Now what ...?

Before we left Nottingham we knew we'd probably need some other form of income, at least from time to time. Going back into teaching did not really appeal, although it was obviously the most efficient way of using our expertise to bring in enough cash. Instead, we drew up a List of Possibilities.

We've always been great list-makers. Nothing quite beats the satisfaction of crossing off a job, particularly if they are done well enough to stay done.

Our lists tend towards the pedantically detailed, thus allowing us to cross more jobs off. 'Fence field' might be split into 'buy fencing gear'; 'dump posts and wire'; 'bang posts in', and so forth.

Really unpopular jobs might get sub-split, with algorithmic intensity: 'buy wire; buy posts; buy staples; distribute posts; distribute wire; lose hammer; spill staples; scream loudly and repeatedly'. You get the drift. (No, I do not enjoy fencing.)

Our List of Possibilities was an exercise in brainstorming; well, on my part, it was. Anne tended to have just one or two patently sensible ideas like 'get a little part-time job', or 'do a bit of tutoring', while I tended more towards the baroque end of the scale, with ideas like these, found lurking in an old notebook:

- adapt the game of Scrabble to a learning tool for students of chemistry, and then make and sell it
- make wooden toys and dolls' houses
- make garden sculpture, from, er, wood, or metal, or something
- design false and amusing pedigrees for dogs, for sale in

souvenir shops (also possibly for cats, budgies, hamsters, lobsters, etc.) Could be very big, this one

- take portraits of people's houses, with or without the owners posed in front
- import US car numberplates to resell via Exchange and Mart
- design slip-on covers for paperbacks, so the reader could mislead the casual observer/parent/etc. Thus the steamy novella might transmute into 'An Advanced City & Guilds Course for Quantum Mechanics'; by corollary, the school swot might gain some cred by being seen reading 'Blood, Sex and Projectile Vomiting; and Afterwards Tea on the Lawn: a Dominatrix Reminisces'
- design Ex Libris sheets, spurious and otherwise for sale as souvenirs (Ex Libris Moses Iudeus. Tabletum Unum)
- breed a strain of elderberry that would challenge the grape for size, flavour, vinosity, elderberrosity, etc.
- ☺ and er, yes: sell garlic-flavoured icecream out of a wrecked tricycle.

All splendid and highly practical ideas, I'm sure you'll agree. The design and printing jobs cropped up because we had inherited a little Adana letterpress machine which we'd had a bit of fun with, and we'd specially bought an old Gestetner to bring down with us, in case it came in handy. I should point out that I knew perfectly well that one cannot do professional five-colour jobs on a Gestetner, but one could make a start, with, say, one colour: black, say.

As for making dolls' houses and so on, this was surely just a matter of getting on with it. My previous woodworking experience was limited to pruning firewood and smashing up a Blenheim Orange, but I knew one end of a hammer from another, so how hard could it be?

My own favourite entry in the diary list was 'play bass guitar in a band in Swansea?' Even with the tentative question mark, this was well outside the realms of reason, I must admit. Yes, I did have a bass guitar, but no, I could not play it. Also, we were going to be 40 miles from Swansea, where I knew not a soul, never mind a band-meister looking for an incompetent rhythm section. Also, I had not adequately appreciated how tired I was going to be after a day's proper grafting, and that a 40 mile drive for a gig that ended in the wee small hours, followed by a 40 mile drive home, with an early start next morning … was not, well, 'on'.

Also … well, let's just put that one to bed, shall we? File it under 'Pretty Desperate'.

# Courgettes, Winter 1989

For the Chinese it has been the year of the Snake. For us, it has been the year of the Courgette.

It started in March and April, with us hacking away at huge bales of compressed Irish peat* before sieving it, garnishing it with dolomite, sharp sand, seaweed meal, blood, fish, bone, sweat and tears, and lightly tossing it in our Polish DIY concrete mixer for thirty seconds or so. We then gently pressed the resultant potting compost into fifteen-to-a-tray Plantpaks, each unit nesting neatly in a standard cheapo seed tray which offers some support to the flimsy plastic. We sowed the seed, carted the trays into a polytunnel, watered generously, and stood well back. In the merest nick of time, I might add, as they all came up like rockets.

Meanwhile, back on the field … we'd spread most of our composted muck onto the courgette patch. The cows and sheep had already grazed off the weeds, and the ducks had thoroughly mugged, splodged, and gulped all available slugs. Then it rained.

When it stopped, I ploughed as shallowly as I could, given the nature of the land and the equipment and my own almost total lack of skill. I tried to turn all the weeds and compost three or four inches under, ready for the exploring roots of the courgettes to find.

Alun, the mechanic up the road, did a Portland Vase job on our venerable spring tine harrow, and I bashed the

---

* This was in the days before people worried about peat. Personally, as long as the peat remains in the horticultural ambit, I can't get too excited, especially as millions of tons of it are burned in power stations as ludicrously inefficient fuel.

ploughed ridges down into a rough seed bed. As an experiment, we laid six 200ft strips of black plastic mulch to bring on some of the plants by a week or so.

We planted out all 3,800 Ambassador and President seedlings at the end of May, as usual, fully expecting the seasonal freezing gale that seems to greet all our semi-hardy plants. But it never arrived.

What is more, as we dug the holes to drop the modules into, we found the tip of the trowel was just scraping into black compost. Perfect. For once, my amateur ploughing had got it spot on. The seedlings were in.

By now we knew we'd already used up our season's luck, and resigned ourselves to a cold wet summer. We greeted the 'dry spell' with scepticism.

We were still sceptical when the spring that supplies our field water dried up completely and without precedent, and we had to revert to the expense of mains.

The courgettes loved the heat. They exploded into bloom and inflated their fruit like those peculiar condomesque balloons that clowns at children's parties twist and knot into dachshunds and giraffes. They fruited faster than we could pick them.

We cut them at six inches long, allowing two days for them to put on another inch before cutting again, but they were having none of that. At the peak of their flush, they were growing an inch a day. We were getting up at five in the morning to pick them, knowing that by noon they would be too big, and even with the help of John, our intrepid WWOOFer,* we

* So what is a WWOOFer exactly? It's someone who spends a period of time working on an organic farm in exchange for their keep, and possibly tuition. WWOOF is a splendid international institution. More details via an s.a.e. to: WWOOF, PO Box 2675, Lewes, East Sussex BN7 1RB. Or e-mail hello@wwoof.org . Or visit www.wwoof.org .

were only just keeping up. By the time we'd finished picking the last row, more fruit in the first row were ready. I've never seen anything like it. Hundreds of pounds of perfectly good, but marginally over-sized prime veg went to the cows. Productivity over last year rose from 140lb in the last week of July, to over half a ton.

It couldn't last, of course. Everybody was over-producing like mad and the bottom fell out of the market. Even the new bulk sales through the supermarkets couldn't cope with the sudden surge. Our packhouse tried chilling and stockpiling against leaner times, but it didn't work, and tons had to be dumped, one way and another. Even my own bright idea of slowing the plants down by allowing some to grow into a single marrow merely produced another weighty problem. Would anybody out there like to make 850lb of chutney? We'd have been better off knotting them into dachshunds and giraffes when they were little.

It's been hard work, but it's been a memorable courgette year. As I write, they are still producing 100lb a week. They've already given us over five tons, I should think. Marvellous and wondrous prolixity.

One major complaint: Did you know that Tesco, for example (in 1989), will not accept perfect courgettes if they are over six inches long? It's because they will not fit their stupid little produce trays. A six inch as opposed to a seven inch courgette represents a hidden price cut to the grower, a misguided sense of aesthetics chez Tesco, and an insult to the consumer.

I appeal to all courgette buyers to contact their local supermarkets and harangue the manager politely but firmly. Tell him that you would like longer courgettes, that you don't mind in the least if they have a slight crook in the neck or are not perfectly uniform in shape, and that when you buy food you intend to eat it and not hang it on the wall to admire the absurdly expensive and wasteful packaging.

Ask him when he intends to stock a full range of organic goods, how soon he will be abandoning excess wrapping in favour of quality controlled loose bins, and when he will stop profiteering and either bring the retail price down, or pay the grower more.*

And while you're at it, ask if he'll be getting in any more of that nice cheap Vin de Plonque he had in last week. Hmmm. Do I spy a paradox looming here?

\* \* \*

Yes … I remember that season well. Largely because of my back …

One of the snags we unexpectedly found ourselves up against when we started working the land was the way we were forced into acting and tooling up as a direct result of the local weather.

Our intention, before we arrived, was to work as lo-tec ('lo-tecly'?) as we could, using simple and non-polluting tools and equipment. I even bought a splendid book called *Pedal Power* which I hoped would give me guidance on how to build a pedal-powered plough. No, don't laugh. In the right soil, with the right climate and geography, it's not impossible, as long as you think 'static power plant' rather than 'Tour de France'. But we didn't really have the right soil (too stony),

---

* This was written in 1989. I believe things have improved a little since then. Courgettes, for example, are now in papier mâché trays, eight inches long or sometimes even sold loose. A definite improvement. I wonder if HDRA members haranguing made any difference?

and geography was against us (too slopey), and so was the weather (too sloppy).

Yes, 'the weather' …

If you want to grow on say, an acre of land, including planting out some thousands of urgently growing seedlings, you need several consecutive dry days to do it in. Two days say, for the land to dry out enough to get onto it, and then however long it takes to cultivate the field; and then the actual planting.

Digging is just far too slow as a means of cultivation. Even a rotavator would take a week to do the job properly, and requires an answer to the question: Are you going to get nine or ten consecutive dry days, which coincide with the optimum plant-out time for several thousand courgettes? Not a chance; not round here, mate. So you will be doomed to having thousands of seedlings getting pot-bound and bolting in their little plastic cradles while you wait for another window of opportunity.

The only practical option is to plough and harrow, which is quick and efficient, but which means you have to have a tractor, and all the grief and time-wasting that machinery brings with it. We didn't *want* a tractor; we simply had to have one.

So: two days for the field to dry off; one to plough; one to harrow; one to rotavate the seedbeds; one to plant out. Five days: not unreasonable. And at the end of them, you have all your urgent seedlings planted out, *and* the remaining three-quarters of an acre ploughed, harrowed, and awaiting your pleasure and attention. There's no alternative, is there?

In The Year of the Courgette we got our timing pretty much right (people tend not to appreciate how much good timing is essential to good farming), but knew we had to get all those 3,800 seedlings out as fast as possible.

It was a big operation, and needed careful planning. What was the most efficient way of shifting more than 250 moderately heavy trays that could not be stacked, up to 100 yards uphill, to be planted (half of them) through non-existent holes in incredibly tough plastic sheeting?

Each one must then be planted into a scooped hole, firmed in and adequately watered, by hose at the bottom end of the field, but by watering-can at the top end, where the hose pressure splutters to zero.

Then all 250 seed trays and 3,800 Plantpaks must be collected and carried off the field, ready for washing and storage.

That's a lot of work. I'm very glad John was with us. (I wonder if *he* was? Must ask him one day.)

The way we finally worked it was to have one person collecting the trays from the tunnel, another distributing them to appropriate points on the field, and cutting carefully-spaced holes in the plastic with one of those wonderful snap-off razor knives, and the third planting and watering in the plants.

Our greatest ally in all this was a splendid double-wheeled trolley* we'd inherited, which would hold eight trays at a time, which was better than carrying them three at a time, waiter-style, but still much too slow to keep delivery speed up with the field workers' pace. BUT! When covered with an old hand-crafted wooden roof-rack off a friend's defunct Reliant, Tebbit would take 22 trays at a time. Now you're talking!

I did the planting, as I reckoned it was the hardest, nail-rippingest task of the three (yes, how noble can you get? But

---

* Named 'Tebbit', for reasons you don't want to know about. Careful examination showed the word 'Fyffes' had at some point been inexpertly hacked off the rubber handles. A story there, methinks.

read on …). We worked and worked, and got nearly three-quarters of the plants in before exhaustion and darkness supervened. A day of triumph.

The next morning I couldn't bend over to pull my trousers on, never mind my socks. Oh, the humiliation. Stiff as the proverbial. So, while the hero of yesterday ooped and aaahhed his way up and down the field, gingerly delivering one tray at a time into the patient hand of his colleagues, Anne and John doubled their work rate, and got the job finished. I continued to oop and aaahh for several more days, as I recall.

*That's* why I remember that season so well …

# Cheap at half the price ...

The actual finding and purchase of Dwlalu Farm was an interesting process. My parents drove down a couple of times in their VW campervan and trawled the estate agents, handing out carefully Gestetnered lists of desiderata and spending days at a time looking at patently unsuitable properties before reporting back to base and awaiting the estate agents' further brochures and leaflets.

What we required were notices of properties of approx three acres of growable land, with double-dwelling capacity, within reasonable reach of a town. By return of post we received a dozen or so inch-thick A4 envelopes stuffed with details of derelict manses, derelict hovels, town houses, modern bungalows with no gdn, but fll ctl htg, htd twl rls, nc gld tps, etc, etc; but amongst all this dross there were five possible possibles.

Mum and Dad drove down again to examine them. Two of them turned out to be flat-pack shacks apparently freestanding on a bed of rock, with a decorative goat nailed to a nearby cliff for effect ('broad pastoral views'); two of them they never found at all, despite map references and Dad's years of experience leading convoys all over the UK in the blackouts of wartime; and one of them looked quite promising.

It was on the outskirts of Carmarthen and fulfilled most of our modest criteria. I jumped on a train to see for myself.

Yes ... pretty good. But somehow ... the house was a bit clunky, and the land was a bit small, and full of reeds, which suggested that expensive drain-laying was a matter of urgency. 'We'll think about it. Thank you.'

By now Mum and Dad had looked at dozens of places, and this was the best they'd seen, by quite a long shot. We began to lose hope of finding anywhere right, and drove the two hundred miles home in moderate silence.

On the mat when we arrived was another small mountain of assorted bumph. All rubbish. 'Desirable top storey attic (no pets, no window-boxes, no visitors after seven)'; 'Charming Norman keep, ready for restoration'; 'Chip shop, with planning permission for bingo hall and airport'.

What is it with estate agents, that they seem to be entirely impervious to what it is you actually want?

'I would like to buy a perfectly ordinary three-bedroom semi within twenty miles of Derby.'

'Certainly sir. I believe Fontainebleau will shortly be on the market, but meanwhile perhaps sir would like to look at these machine-washable wigwams?'

But then ... next to the bottom of the pile ... an A4 envelope with just one sheet in it. It described a place that fulfilled all our requirements except the double-dwelling, which we knew was always going to be a problem. We grabbed a few hours' sleep and all six of us set off back to Wales.

And yes, it actually did live up to its description: a nominal ten acres, including woodland, driveway and yard; plenty of outbuildings; two good fields; and a house that was big enough to sleep us all. No real double-dwelling possibility, however. Never mind: this was clearly the one. We could work on the living-space later.

I did a bit of rudimentary soil-testing and we poked around and asked questions. Yes ... this was far and away the best proposition we had seen so far or were ever likely to get. We said our thank-yous and set off for the nearest pub, where we sat round a big rickety table, not wanting to be the first to speak.

'Well I quite liked it.'

'Mmm, yes.'

'What about you?'

'Oh, er ... nice view.'

'Mmm. What was the soil like?'

'Bit shallow, but looked OK.'

'Lots of handy out-buildings, I thought.'

Eventually someone cracked and said 'This is the one, isn't it?' and we all piled in and agreed.

What is more, it was cheaper than the other one we'd looked at. A smallholding, if you could find one, seemed to be up for £45,000.* This was £39,000.

'Well ... shall we make them an offer?' Dad suggested.

It took no longer than a few seconds to decide unanimously that this one was worth the full price. We didn't want to run the risk of losing it.

So we drove straight back with the full offer. It was accepted on the spot, and the whole deal went through as smoothly as a greased weasel. Within a couple of months we were off ...

---

* How times change ...

# Crop News 1989, Spring 1990

'Nineteen eighty nine wasn't too bad a year, I suppose.' That's about as close as you're ever likely to get to a shriek of delirious delight from a smallholder or a farmer, because, by definition, things can only *ever* be as good as a farmer plans for. Any deviation must be towards the worse: either glut or ruin. And as nothing *ever* goes properly to plan, every season is bound to be something of a disappointment, one way or another.

This, of course, is why farmers are notorious moaners. I once saw a lovely cartoon of a farmer and his friend surveying a field packed with huge healthy haystacks, with the farmer saying 'If this lot catches fire, it'll be the worst year I ever had'. I understand, brother.

Our courgettes did well for us, despite the ironic bad patch of good weather when over-production meant everyone was working overtime for nothing.

In fact this bad patch was something of a blessing in disguise as it led to me dealing with my local wholesaler, Eric, for the first time. I wondered if he might be able to shift a couple of boxes of outgrades for us; please?

He could indeed, and was very pleased to do so, throughout the season, thus enabling us to sell thousands of perfectly good fruits that would otherwise have been rejected by our organic packhouse wholesaler and would have had to go to a spoilt and undeserving cow, who was already falling significantly behind in her grass-trimming duties, due to being regularly and endlessly stuffed with courgettes.

The radishes weren't too bad, either. Well, actually, during the hot summer, they were a totally unmitigated disaster. Nasty, hard, knobbly little purple things with a bite like a napalmed chilli. But the later plantings, from the irrigated polytunnels, did very well. Bright red, round, neat, crisp and easy to trim.

Our other main crop, of Jerusalem artichokes, also earned its keep … just. We have two strips of nine rows each: one runs 200ft from top to bottom of the south-west edge of the vegetable patch to act as a windbreak for the courgettes; the second strip runs down the middle of the patch to keep the gales off the runner beans. During the winter months we dig the artichokes and sell (a few of) them. It's a lovely job when the sun shines, like lifting the spuds, but quite ghastly in a sleety January squall.

## Mystery Scarecrow Turns Out
## to be Frozen Arti Grower
### 'I Was Wondering Where He'd Got To,'
### Wife Admits.

On the whole, then, we're pretty happy with our three main crops, though I think we'll be cutting the courgettes back a bit for 1990, as we can't afford to get caught up in gluts. We'll replace about 500 of them with a selected marrow that we hope we can sell fairly easily. The co-op thinks we can, too.

We're almost certainly going to abandon salad crops completely. Little Gem lettuce, which is in huge demand for the supermarkets, is only worth growing in vast numbers, and we haven't got the space, the labour, or the harvesting and handling facilities. They are also prone to bolting quicker than those paper palm trees they make on Blue Peter if your timing is half a day out. Not for us.

The radicchio, despite the pictures in the seed catalogue,

are just wildly unpredictable: large, small, smooth, deckle-edged or corrugated by turn. Not what supermarkets want at all, and largely incomprehensible to local greengrocers.

'Who wants a brown lettuce?'

'It's red; it's a radicchio.'

'What's wrong with green?'

So what's the point?

Instead of salads we'll probably extend the carrots and leeks. We'll have to be careful with quantities, however. We could easily grow more than we could sell, although the co-op is shifting more and more of both. Our problem is that being small fry, we come bottom of the pecking order.

Salad onions are the other White Hope for the '90s. We grew a double 200ft row of Ishikura this year and sold them for over £40. This was all to retailers, so we got a good price. If we grow five times as many, say, we'll have to sell to a wholesaler and will be paid less per unit, and may get into packaging complications. But salad onions definitely look interesting. Cautious optimism.

The only other field plan we have for the new season is to grow thirty 200ft rows of Maris Bard field beans. The aim is to (1) fix nitrogen for the following season, (2) harvest lots of nutritious beans to feed the cows over winter, and (3) chop the haulms to use as mulch, perhaps for the comfrey. It sounds like a great idea so far. I wonder if it'll work? Watch this space.

On the animal front, we hope the lugubrious April will have a healthy Limousin calf and that Choccie will produce her first bright-eyed Angus. The sheep, all five of them, should be producing seven or eight Suffolk-cross lambs. That's if they can be bothered. We're thinking of buying a proper pedigree

Suffolk ewe this year to begin replacing one or two of the clapped-out old bicycles we have in our present flock.*

Anne has dug a little pond for the ducks in the back garden. It's only small: more of a bidet, really, but they seem to like it, especially the scaled-down diving-boards. To see a Harlequin drake perform a perfect double back somersault, with pike and tuck, makes all the effort of construction worth while. Anne says we should film it for You've Been Framed.

With any luck the fox won't kill any more of them so close to the house, and we can start rebuilding the flock, or herd or whatever. Personally, I think the collective noun for ducks ought to be 'a squackle'. Any offers?

A final animal venture might be to reintroduce geese into the orchard. The last trio did a grand job of keeping the grass down. We lost them on Christmas day two years ago because we were half an hour too late in putting them to bed. The fox beat us to it. The good news is that two of them were still warm. They went into the freezer for next year. It's an ill wind, eh?

I wonder what fox tastes like?

* * *

And I wonder that happened to all those Maris Bard beans? I guess they can't have been a brilliant success, or I'd have remembered. Worth a try though.

---

* When originally published, this description brought an abusive letter, accusing us of cruelty to animals. Can't see the logical connection myself: with increasing age, sheep lose condition, as do people; and no sheep *ever* gets to be a bony pensioner on a conventional farm.

# First things first ... and second ... and ...

We'd done an awful lot of thinking about priorities before we jumped in at the deep end, and now we had to see if our plan of action was a runner or not. We already had a mission statement, a full decade before yuppified corporations caught on to them. In fact we had several, but could never decide which one should head our understated monochrome Gestetnered company stationery. The choice was tough:

- 'Growing Stuff for People'
- 'Veg R Us'
- 'The Best Courgette a Man Can Get'
- 'Peas in Our Thyme'
- '24 Carrots, Inc.'

Top of our priority list was 'No Mortgage'. We knew we'd be taking an income cut of about 75 per cent, which would, we estimated, leave us just about enough income to pay 'real world' bills like electricity and rates and petrol and so forth, but there would be none to spare for the usurers, or even the taxman in the short term.

We also knew how lucky we were that my parents had come in with us, otherwise our No Mortgage policy would have bought us only the meanest of dwellings and opportunities.

Very Low Income meant we needed to get cracking immediately on Very High Output. What should come first?

The Master Plan* was quite sure that the deepest layer of organisation should come first. This was just common sense,

---

* ... and 'Mistress Plan'; we are an equal opportunities non-employer.

31

otherwise we'd be constantly fiddling and adjusting and updating later on: wasting time.

So what were these deepest layers? We thought a polytunnel should come first, so we could grow at least some food for our six selves while we got the fields allocated and cultivated. A polytunnel would need watering facilities, so designing an irrigation system came next.

Pretty well as soon as we'd unpacked, we chose the spot for a 30 foot tunnel and set about finding one. It arrived in Llanelli in a veg lorry from Birmingham, and we picked it up in Gloria the Transit. The weather was fine and still for a couple of days, and four of us got the beast erected with no major problems, apart from somehow losing a hammer in the foundations, presumably in the same way that more primitive cultures tended to lose the skull of their firstborn.

It was our first attempt at a tunnel, and we were pretty pleased with it, and were also pleased with the door system we designed for greatest flexibility of ventilation for the 435sq ft of growing space.

The irrigation went well too. The big pipe that brings our spring water 200 yards downhill from the little header tank enters the house through the bathroom wall. It was surprisingly easy to bifurcate it, using a brass thing with two nuts on and a length of black PVC pipe we found in a shed, and then to lead the new pipe off to a hozelocked watering pipe, which could be sub-split into various directions whenever we felt the need.

So … two easy victories. Smallholding was going to be a doddle.

\* \* \*

Meanwhile, my parents (Ralph and Jessie) had bought a trio of Holland Blue hens, who were supplying us with the occasional egg. The three gaudy bantams we'd brought down with us never seemed to do anything apart from strutting their endlessly funky stuff. But then, they were all males.

Our new neighbour, Ken, came round one evening with a couple of little Mallardy beaks poking out from under his jacket as a 'welcome' present: our first ducks. The start of a love affair that never died.

The final triumph of the first year was that we bought a rotavator. We'd previously put our hopes in a fat bloke with big hair, and his long angular son who, between them, ran a sort of mechanics' bric-a-brac stall in the Nottingham Cattle Market. Their trinkets ranged from venerable tillers and pumps to rusty and interestingly twisted spanners and gimlet-oriented screwdrivers. We told them we were looking for a Howard Gem to use in Wales. 'Ooh … lovely machine. Triple by-pass underhand camshaft. Wonderful stuff. Yes, oh yes, we've got one. Doing it up just now, as it happens. Oh yes. We could deliver it OK. Drive it down overnight. Bit of fun. No problem.' We shook hands on it and I bought a broken Mole grip as a sign of good faith. Needless to say, the Gem never materialised; and the Mole grip subsequently turned out to be utterly useless against moles.

But a local dealer had a very tidy Howard 352 for sale at a reasonable price. Although we didn't realise it at the time, the 352 was a much more appropriate machine for our situation.

For a start it has a tiny turning circle if you're nippy on your toes. In theory, you should disengage the blades before attempting a turn, but hey! life's too short! What you actually do is: disengage clutch handle; tilt forward, and haul the machine round through 180°; swing off-set handlebars from rhs to lhs; re-engage clutch; and you're ready to go … Easy. If you're agile and confident enough with this movement, the blades never have a chance to get into trouble because they're thrashing around in mid-air. Next: tilt back and dig in; tip the machine over; mash calves and thighs to jelly; twang knee-caps into orbit; etc, etc (those final bits are optional).

And if you're *really* proficient, you don't need to disengage the clutch until after the 180° swing. You may even find time for a double entrechat en route between the de-clutch and the handlebar cross-over. I frequently did just so, just for the hell of it, to the massed delight of the sheep and cattle.

By contrast, the Gem has all the elfin manoeuvrability of a supertanker.

The following spring, we put Stage II of the Master Plan into operation. It was in two parts:

A) Get field-cropping underway;
B) Get dairy products producer.

Part A) was pretty easy, as it turned out. Another kind-hearted neighbour, John, drove his big Fordson round and ploughed and harrowed a 200 x 50 foot strip of the top field for us. We stuffed in the garlic and a few assorted veg for the family, as well as a couple of experiments like asparagus peas and Egyptian tree onions. This first season was to be a test of local conditions more than anything.

Part B) of the Master Plan took a little longer to work out. The first choice to be made was 'cow'? Or 'goat'?

What little we'd seen of goats led us to the scientific

understanding that they are, on the plus side, small but efficient producers of modest quantities of milk, entirely adequate for the needs of a smallholding family, and, on the negative side, they are clearly the spawn of Satan; stink; perform obscene acts upon themselves (I've seen them); and have been entirely responsible for the desertification of Arabia, the Sahara, the Kalahari, the Gobi, and most of the spaces in between. Their milk, butter and cheese also stink, of something between fermenting skunk and quintessesnce of corruption.*

A tough choice, but we made it. Next: Dexter or not-Dexter? Dexters are to cows what Ronnie Corbett is to comedians: 'amusingly small'. Quite adequate, again, but inclined to petulance apparently, and anyway, hard to find (and not just because they're so small).

We talked to Ken and Doreen next-door and they said 'Jersey', and explained why. They are big enough but not too big; give more than adequate quantities of the highest-quality milk; don't need firing up on expensive concentrates; are placid and friendly (except for the bulls who are ferocious and malefic); and ... are heart-meltingly beautiful. Unlike goats, say.

So one morning, Anne went off with Ken and returned with 'Daisy' in the back of the trailer, bought from a nearby farm. She was enormous, horned, and opinionated, and didn't much like the look of me. I could tell.

But within a week she became modestly-sized, amenable within limits, and prepared to be reasonably friendly. But still

---

* You may think otherwise, of course.

horned. We were going to get on, albeit at a respectful distance, to be decided by her, not me.

So, our first season was a modest success. We'd got a tunnel up, some watering facilities, some ploughed land, a rotavator and a cow.

On the downside, we were shocked to see brown blotches on the leaves of our beloved Desiree potatoes … blight. We'd never so much as heard of it in Nottingham. We were soon to find that no matter what we did, round here blight is endemic.

# Sheep and Lambing, Summer 1990

By the time you read this, we'll all be well into sowing and summer will be a credible prospect.

But as I write, a gale is battering sleet into every nook and cranny, and I think inevitably of sheep. One in particular: an old Jacob-Suffolk cross called Rocket, daft as a brush but quite amiable, and whom we have just this minute buried in the orchard along with one of her lambs.

Smallholders tend to have mixed views about sheep. Certainly they do a good job following the cows round and keeping the grass in good fettle. They also spread their contributions more evenly than a cow does, but having said that, you've said it all. They'll find the smallest gap in a fence, jump, panic, stand on their back legs to browse your best young fruit trees down to blasted stumps, catch foot rot, need shearing, get riddled with maggots, need dipping, get involved with quite unnecessarily complicated lambing problems, need dipping again, and, after all that, die at the drop of a hat, after they've first had a crack at eating it.

We lost our largest lamb last year. It had poked its head through the pig-mesh (they always do it, even if the grass is greener on their own side) and didn't know about going backwards. By the time we realised what was going on and had released him, he had made his decision. Whatever we did after that was a waste of time. We took him indoors and gave him warm milk and a hot water bottle. I even read him a story about a dear little lamb who escaped from certain death by learning about going backwards.

Not a chance. He'd made his mind up, and die he did. I sometimes think it's fair to say that the only truly successful sheep is a dead one.

Rocket did have some excuse though, as she caught pneumonia. She and her three cronies had been eating off the weeds and sundries in the veg patch by day, and coming into an empty polytunnel by night.* One morning Rocket looked ill and sat down.

Ken next door suggested it might be twin-lamb disease, akin to milk fever, which could be fatal. She was fading fast. We then had a difficult decision.

Should we call out the vet? He would charge £20 call-out fee, plus parts and labour *and* VAT, and a sheep is only worth £30. On the other hand, we might be able to save her lamb.

Yes, we called the vet. He gave her the magic calcium and a couple of jabs. She rallied, then went under again. Anne managed to release the lambs, but it was too much for the mother. There were three lambs, in fact, all intertwined in an obstetric knot that only a sheep could have contrived. The ram lamb was small but full of life and bounce. The two ewes were a bit slow and feeble-looking.

Would you care to guess which one died that evening, against all logic?

That left two ewes. We fed them some colostrum couriered in from one of Ken's sheep and tucked them up in a warm box on top of the stove. Next morning they were still there if still feeble. We weaned them onto Denkavit replacer milk and kept them warm. Then suddenly the larger one went into collapse. Its eyes closed and its breath became very faint.

Pneumonia.

---

* Two months ago we had five sheep, but one left her mother behind and ran away, scrambling over a low wall and two fences to join the flock next door. Extraordinary. We can only assume that she must have struck up a meaningful friendship during the tupping season.

Anne massaged its chest and poked its nostrils but it continued to fade. She gave it a tiny jab of its mum's antibiotic but nothing happened. That left one. It fell to me to break the news to Caity. She wept a few tears but soon got over it. She knows about sheep.

Blow me down, three hours later the blooming thing was back on its feet, bawling for its dinner. The antibiotic must have done the trick after all. Sheep will never cease to amaze me.

PS. I only once genuinely outwitted a sheep. We had a particularly devious and grumpy ram on a chain in the orchard, keeping the grass down. Every morning I took him his breakfast and moved his tether to a fresh patch. Every morning he would swing round suddenly and try to deliver a bone-cracking butt to my kneecaps.

One morning I took a sledgehammer with me. I can still see the look of profound respect in his eyes as he swung round as usual and hurled his full weight into a stationary 14lb block of iron. We got on much better after that.

\* \* \*

I can still hear that hammerhead ringing out. A slightly sharp B flat, as I recall.

# Heating the breeze ...

We spent our second winter doing much the same as we'd done in the first winter: trying to sort the house out, and keep warm.

The building is basically a traditional Welsh two-up-two-down small farmhouse, built circa 1695, a local historian thinks, which has been tagged on to a bit since. The nineteenth-century dairy building has been incorporated and become the kitchen, and we are blessed with a little extra pantry or larder, which has now become the utility room.

Tagged on to both of these is a 1960s flat-roofed bathroom whose interior walls were pixelled with swathes and galaxies of black mould when we arrived.

Tagged on to the kitchen is a moderately-sized '60s extension that included the back door and the stove. This became the dining-room cum transit station for stuff being shifted in and out of the house. We stuck up some steel shelving to hold racks for coats and secateurs and boots and gloves.

Wellies were allocated their own strip of newspaper, and everyone had their nominated space to which they must stick. Twelve wellies scattered about would mean a broken leg for someone sooner or later, probably while carrying a box full of eggs, surgical scissors, and a double handful of loose razorblades.

It was hard to know where to start with the house. It wasn't exactly filthy or dilapidated, but just generally run down and scruffy. Everything needed painting, except for an extraordinary slab of lath and hardboard – some of it gaily perforated – that covered an entire wall of the sitting-room. This masterpiece of DIY concealed the fractured and battered remains of a hideous oxblood glazed-brick stove, and also fronted a couple of built-in cupboards. But the whole absurd construction was so cheap and makeshift that the cupboards were virtually useless. At over six feet high, with only one rudimentary shelf each, these 'cupboards' were actually suitable only for storing brooms and models of giraffes.

And the whole mess was thickly painted an incredible, and unsuitable, and depressing pea green: the precise tone that was used to add a touch of Parisian gaiety to some of the grimmer sanatoria of the 1920s.

It seems to be a personal idiosyncrasy, but I find green a very difficult colour. It looks great on fields and parrots, but all wrong on woodwork and electric guitars. That green, and the hardboard, would have to go before I lost the will to live.

But there were higher priorities. The topmost was the end room downstairs, which was the obvious bedroom for my parents. It was damp almost to the point of clammy. The only source for the damp that we could think of was the chimney on the gable end ('pine end' round here) which copped the full weight of the Atlantic rains. It would have to be felled.

Accordingly, two lads, Dave and Julian, turned up and amazed us with a scaffoldless steeplejack tour de force. Gosh, they were quick.

Their work didn't completely solve the damp problem though. Some of it must have been rising, which is not surprising in a three-hundred-year-old house made of rubble and mud, with no trace of a damp-proof course. I was surprised at how relatively dry the place was, considering. Still am.

Upstairs, we had a big bedroom for us, a smaller one for Paddy, and a tiny boxy little room, which had been skilfully conjured out of part of our room and the stairhead. This would do fine for Caity, whose requirements were still only those of a five-year-old: ten acres to roar round in all day, and a small box to crash out in all night. Her nick-name was Aurora, actually. Nothing to do with the mystical light of dawn, but all to do with her being 'a roarer'.

By now Paddy was attending his fourth school in as many years. He'd left his Nottingham Primary, joined the local Primary, sat the 11+ exam (still running here then), moved to the local Grammar school for a year, then finally moved to a local Comprehensive, following reorganisation. The experience didn't seem to bother him too much, and he was settling in and enjoying it.

Paddy's nickname was Egon. As in 'where's *he* gone?' as soon as you turned your back on him in a shop.

Cait, meanwhile, had joined the nursery group as soon as we arrived, and went into the Primary (a five minute walk away) shortly after she was four. We'd caught that window of opportunity pretty well.

\* \* \*

The most important Green Job was to check the loft insulation. Access was not easy, as the hatch was about a foot square, and I am not. But somehow we levered me up and in, rather like forcing a pig through a cat-flap, as nobody pointed out at the time, but I know what they were thinking.

The only form of insulation, as we expected, was a thick layer of dust, enlivened here and there with bits of plaster, dehydrated rats, and, oddly, an ex-pigeon. How on earth did it get in?

All our expensive heat was going straight out to heat the winds of Dyfed.

We applied for a suitable grant and, rather to my surprise, got it. But it covered the cost of only four inches of insulator. Clearly this was not enough, so we bought enough rolls to lay six inches, which Green People seemed to think was a bare minimum.

It was the horriblest job I'd done so far, and that included taking the assorted clowns and hooligans of Mechanics III Day Release for General Studies, on the graveyard shift, last thing on a Friday afternoon when everybody else had gone home. Oh, yes, how well I remember the day they unscrewed the bowl of my chair from the legs, so that the bowl and I fell off backwards onto the parquet. Oh, how we laughed!

The worst problem, apart from the catacomb dust (is that *really* a row of dead popes over there?) was lighting the job. True, I did have a new fluorescent inspection-pit light, but it was on a pitifully short leash and somehow managed always to be either shining straight into my eyes, or be hanging awkwardly between me and the work, or be directly behind me, casting cartoon shadows over the bales of fibreglass. But usually, it was just ten feet too far away, and far too dim in the first place.

And then there was the lack of treading-space. The roof had been put up in an age unhampered by any form of Building Regulation, and the slender joists were well over six feet apart (two metres plus). Between them there was only the fibreboard that doubled as a ceiling when viewed from below. How could I manoeuvre a six foot gap on my hands and knees? And how could I get into the eaves area at all? We needed to bridge those joists. Luckily, we had brought a couple of eight foot planks with us, from Cait's old sandpit.

The first job was getting the planks through the hole. That alone took half an hour, as they needed trimming to go in at all, and the trimming meant it was then touch and go whether they'd be long enough for the job.

Then we couldn't get the unopened fibreglass rolls through the tiny hole either, and had to open the bags and feed the horrible stuff up, naked, as it were. We wore gloves and hats and face masks, but that glass fibre gets everywhere. Yes, including there. Just don't wriggle.

It was cramped, it was hot, it was constantly dusty, it was stuffy behind the mask, and my glasses kept fogging up. I couldn't keep a light on my work, and kept losing the scissors; that is, when I could prise them off my gloved fingers. Eventually I gave up and used a razor knife, which cut better, but which got lost even quicker under the huge cumulus heaps and billows that piled up everywhere. Every half hour I forced myself back out of the hole and had a pint of home-brew.

By the time I'd finished the job, I didn't know what my name was, and didn't care. The job was done … and I hadn't once stuck my foot through the ceiling. Amazing, but true.

I never want to go into that loft again.

# Crop News 1990, Spring 1991

It's not been a very good year, we have to admit to ourselves. For a start, the overall price for our courgettes was down on last year, despite a rising market. What is even more puzzling is that our top-grade Supreme Classic Designer SuperStud Turbo courgettes, as supplied to the organically-conscious supermarkets, fetched in August 9p a pound, while all the 'outgrades', including slightly large and even slightly bent ones, dumped on the conventional market for us by Eric, the local wholesaler, as Class 2 Ordinaries, fetched over 20p a pound. How can this be? And can we afford to grow Organic Class I fruit, when Inorganic Class 2's are worth a great deal more?

Our next problem was with the Ishikura Bunching Onions. The trial row we grew last year behaved perfectly, but this year with similar weather to that of '89, they declined to co-operate. The first bed produced well, but just as the second bed came into flow, leek rust struck. We had to rotavate in some 15,000 plants: about 90 per cent of the crop. But the onions hadn't finished with us yet. More of this below.

The radishes behaved much as last year – they hated the dry weather and were largely unusable. More or less a dead loss, as well as the onions.

The marrows cropped reasonably well but the price we got for them was poor, and they are so heavy! It's no problem collecting one for dinner from the back garden, but carrying 80 or 90 anything up to 200ft to the barrow is surprisingly tiring. And then you have to push the barrow!

Again, we had a lot of trouble getting the marrows to live up to the supermarkets' expectations of what a marrow ought to be. The plants became quite truculent and then refused point blank to grow the required dark green and off-white striped fruit to the precise length $x$, diameter $y$ and weight $z$, as stipulated be some wretched pin-striped committee. I sometimes wonder if any of the Quality Parameter Executives of the supermarkets have ever seen a plant growing, much less tried to grow one themselves. I guess they must have nothing better to do all day than make up silly 'standards'.

'I know … how about "Two feet long, an inch and a half wide … with 'marrow' running right the way through it in … pink, no, *orange* lettering"?'

'Hmm. Jolly good idea, Giles. See to it, will you, Annabel? On to a winner there, I think.'

The Jerusalem artichokes cropped beautifully, as ever, but, as ever, nobody's terribly keen on buying them. This also is a great mystery as I think they're terrific, especially in a soup with tomatoes. 'They' say artichokes give you wind, but I can't say I've noticed. Perhaps we could market them as a practical aid to levitation.

Our last major crop, carrots, grew very well. We planted eight rows to a 48in bed and flame weeded after about ten days. The flaming worked impressively (we left a small patch unflamed to compare with) and a couple of hand weedings done later ensured a fine crop.

However … the carrot fly has struck very early and heavily and at the present time we are rejecting over 50 per cent of the crop. And it can only get worse in the coming weeks. The cows love the carrots though.

So, overall, a bad season.

But no season is ever a total disaster. Lemmie, our Limousin-cross calf is already as big as his Jersey mum, and looks set to get huge. The sheep are looking well, and we've bought in a trio of Southdowns to go with them, mainly for the wool. We now have ten ewes and may qualify for the £8 a head subsidy that keeps all those proper sheep farmers in Range Rovers.*

The other good news is that the garlic planted through holes in old silage bags has grown well. No weed problems of course, and the extra warmth retained by the blackness seems to have helped with bulking. We'll be repeating the experiment next year.

Finally, in conjunction with HDRA, we ran a few green manure tests, with strips of trefoil, two different clovers, phacelia and mustard. In our view, mustard won hands down, with phacelia a promising second, as long as you don't let it grow too stemmy, in which case it clogs up the rotors of the vator. The legumes all took too long to establish and became very weedy, thus actually making the weed seed problem worse.

And really finally: remember the onions? After rotavating thoroughly three times we sowed tares to overwinter. In a bare patch they have thrived, but in the onion beds there is scarcely a sign of tares, though quite a lot of truncated onions. Why is this? Allopathic exudations? I suspect so. The onions have definitely had the last laugh.

We look forward to 1991.

\* \* \*

Those Ishikuras were too good to be true. Pity about the Red Pest. They grew really well, and there was a guaranteed

---

* I seriously believed this at the time of writing. I hadn't yet realised that subsidies are only available to big farmers: the ones who are rich enough to have the time to sit on the committees that make recommendations on things like, oh, 'subsidies' for example.

demand. What's more, they didn't need too much post-production work, or 'preparation for sale'.

This is a very important issue for a grower. If he can grow two crops of equivalent value on an equivalent space, and one of them can be chucked straight into a box while the other requires sorting by size, then washing, trimming, drying, and gently laying between sheets of paper on trays … which one is he going to go for? It's not a question of greed, here. The unit price to the grower is a small fraction of what you pay in the supermarket (guess who gets the rest?), but it is the grower alone who runs all the risks of weather, pest, and being let down by wholesaler or retailer. He *needs* some high value/low work crops if he's going to keep going. Unfortunately, there aren't many of these nice little earners.

And should a new one show up, it is rapidly appropriated by the Big Boys who can pay casual labour slave wages and thus benefit from what we call 'the economies of scale'.

The grower also needs to spread his risks, by not relying too much on one crop, for all the reasons mentioned above, while all the time working within the parameters of crop rotation demanded by the organic system. Quite a complex and tricky business, planning crops for the next year.

# You dirty rat ...

Those heavy marrows were nothing compared to the pumpkins we grew in our second season. Our neighbour-but-one, John, had helped us collect a trailer load of free horse muck from a local stable in danger of being swamped by the stuff.* He dumped it in a neat heap and left us to it. We had no means of shifting it around at that point and just left it to mature in situ over winter.

Come the spring the muck had rotted down beautifully and was looking very good. But we still couldn't sensibly shift it onto the growing patch. We considered filling our trouser pockets with it then letting it slowly dribble out in selected spots, like they did in *The Great Escape*, but it would take so *long*. So we began to think about how *really* useful a tractor would be. Hmm.

Meanwhile, it would be silly to waste all that growing-power, so we did the next best thing, and studded the heap with four rumbustious pumpkin seedlings. By the autumn, we had a dozen enormous pumpkins, most of them weighing well over 50lb.

Very good. Now what? What do you do with one huge pumpkin, never mind a dozen of them?

By now we were selling a fair bit of stuff via two outlets: a big hotel in Newcastle Emlyn, and the WI market, just across the street from it. We sawed a couple of pumpkins up into 1–2lb sections, and packed them in pastry bags – half paper/ half cellophane – and sold a surprising number

---

* 'Oh, Mr Archer! Come quick! Eddie Grundy's just going down for the third time! Sorry? What? Oh … Thank you kindly. A small sherry, if I might, Mr Archer.'

at the WI. The sale to the hotel, however, was to prove embarrassing.

The normal deal was that I'd show up with the order from the previous week, tell the chef what we'd be able to supply next week, then bring that order; and so on. It was a sensible system and it worked well for both parties. They got top-notch stuff, and we got better than wholesale price.

'Do you want any pumpkin?' I asked the chef. 'We've got a fifty pounder ready next week.' 'Oh yes,' she said. 'I could make some soup. How much?' 'How about 10p a pound?' I suggested. This sounded reasonable to us both, so I supplied one.

Two weeks later I was confronted by the under-manager who was full of bile, accusing me of ripping them off. He didn't seem interested in the fact that his agent, the chef, had agreed to the price. The problem was that he thought £5 was extortionate for something you cut holes in and illuminate with a candle in the porch for Halloween.

Well, yes … but I wasn't selling decorative pagan hotel furniture; I was selling prime quality organic food to make soup with, as agreed.

No dice. Copybook blotted; never darken doorstep or towels again; cheat; all-but-thief.

It was a horrible feeling, to be so accused. And to lose a very useful customer.

Never mind; life goes on. We never grew lots of pumpkins again though.

\* \* \*

Porky loved the place. Well, what dog wouldn't? Two great big fields to wander about in, with tantalising odours of rabbit and fox, badger and weasel, buzzard and polecat.

We'd bought Porky from an animal rescue home about five years previously. The owner told us over the phone that she was half Staffordshire bull terrier and half whippet (I correctly assumed she meant the dog) and was a real cutie. And so she was: pure white, with black eye-liner, nose, and lips, one pure black ear, and a big black splodge on her head where the other ear's colouring had missed its target. A bad Friday at the Puppy Finishing Shop, methinks.

'Goin clubbin tonight?'

'Yeah. Gorra get me hair done. Wish that flippin whistle would hurry up.'

'That's it.'

'Right. I'm off.'

'What about that last ear?'

'Oh bother … there. Oh blast. Missed. Never mind, no one will notice.'

*We* noticed, but we didn't mind. She looked lovely.

Porky did her bit for us all, keeping the fox away from the ducks, but she was inclined to wander if you took your eye off her, and in sheep country 'wandering' means 'early death from shotgun wounds'. This meant we had to keep her tied up rather a lot, so she didn't catch many pests, like foxes and rabbits, for example.

Our main catchers were a couple of surplus kittens we were given. The kids named them Pudding and Twinkle-Ears (no, I don't know, either). They both excelled at catching mice and voles and, alas, occasionally garden birds (but never pigeons or magpies, the ones we'd like to see caught). Their speciality was catching baby rabbits, and being beastly to them.

They ate most of the baby rabbits they caught. Well, strictly speaking 'half-ate'. Two or three times a week during the spring we would find a pair of miniature furry trousers, on the path or by the compost, or on the step. It was like some sort of grisly Mafia game for cats, played with rabbit trousers rather than horses' heads.

They stayed with us for years. Then one day Pudding went missing. We found him months later, a mummified and flattened sandwich between (not under) two bales of hay. Heaven knows how he came to be there. Perhaps he was set upon by a gang of rabbits one night, and lured ...

Twinkle-Ears lasted much longer, eventually succumbing to one of those endless wasting conditions cats seem to get.

But for all their prowess with small furry things, they never caught a rat. That distinction went to Porky. One afternoon, while rummaging round the strawberry patch, she suddenly stiffened, and pounced, and yelped. A five-minute duel ensued; a little neolithic scene repeated before our eyes, with urgent squeals and growls. Then a final surge, and silence, as she bent forward into the depths of the foliage and carefully lifted a foot-long rat. Her nose was running red with blood, but she was unmistakably triumphant. Her eyes were alight.

'Look what I done, boss.'

'You're a hero, Porky.'

And so she was. I for one wouldn't fancy taking on a big rat, least of all with my bare teeth.

# Crop News 1991, Winter 1991

Last year we grew six beds of carrots, each 200ft by 4ft, with up to eight rows per bed. It was a serious issue, weeding nearly two miles of carrot row, but an initial flaming helped a lot. The crop was good.

This year, because we've had a continuing health problem, we decided to cut back by 50 per cent overall, and by two thirds on the carrots. Two beds would still be weedable, where six would be excessive if only one of us was available to work.

That was the impeccable theory. The peccable practice was otherwise.

The seed went in fine. We waited ten days or so and looked for the first signs of germination, as usual. Then we flamed the beds, on the principle that you flame and kill all the weeds that have germinated just a day or so sooner than the carrots, thus giving the carrots a clear start before the next flush of weeds wake up.

Then it rained.

Sounds great so far, but it had been very dry for the previous four weeks, so the carrots were not as near to sprouting as we thought. Just one or two in sheltered crevices had poked through; and they were the ones we'd spotted.

Our flaming had been seriously mis-timed. This meant that the carrots, when they did finally start to sprout, had to slug it out with all those professionals out there … the weeds.

The end result was that within a week of the first rain, the two beds were ablaze with groundsel and assorted vetch. We were able to rescue only one bed before eight weeks of rain submerged everything in a rats-nest of chickweed and creeping twazzocks.

The courgettes suffered too. We planted them out at the end of May, just in time for the cold rains. The plants couldn't get away, and stayed close to the ground where they were scythed down by the massed ranks of slugs hiding in the weeds and under the black plastic mulch, licking their nasty little radulas in anticipation. We lost a third of the plants.

And we'd already lost a thousand plants in the tunnel before they even germinated. Whole trays were burgled by mice, who piled up little heaps of empty seed husks for us to admire. We trapped two, and, amazingly, I shot a third one.

### Farmer Takes Law into Own Hands: Hold it Right There, Mickey!

Even the green manuring schedule was a disaster due to the weather. In fact the only crop to thrive as well as ever was the row of Hurst Greenshaft peas. In nine years we've never had a crop failure with them, and never had even one damaged pea.*
I wonder what we're doing right?

April the Cow produced another beautiful Limousin calf with no apparent effort again. We noticed she'd started labour,

---

* This period now extends to 19 years with not a single damaged pea.

came for a cup of tea, and by the time we'd had it, so had she. Healthy animals, unforced and uncrowded, do normal things normally. They rarely need vets.

The sheep have been splendid. Our two Southdown ewes each produced twin lamb teddy bears, each more charming than the last. Does anybody out there need one?

The rest of the harem also did their duty, more or less, although the Leicester Longwools seem to have been too fleet for the Southdown ram. He ain't built to sprint.

So we're ready for the winter now. It's been a disappointing year, but it's finally over. We've got the spuds in, bought in the hay, straw and coal, and are preparing to batten down the hatches against the deluge, to plan next year's victory that will make up for this year's whitewashing.

The essential job is to muck out the cowshed … which means mending the muckspreader … which means shifting it … which means getting the tractor going … which means charging the battery … which means finding the battery charger …

Piece of cake.

### Ram

*Wham bam thankyou mam*
*and on to the next.*
*No nonsense over 'love'*
*or 'everlasting hearts entwined'*
*or 'there will never be another ewe'.*
*Just wham bam ram bam wham.*

# The image of his mum ...

We really got down to it in 1984. The house was just about inhabitable, if not startlingly beautified, and the kids were successfully plugged into the educational system, and happy. We had bought another 30 foot tunnel in the summer of '83 and had successfully adapted the irrigation system, and were now producing plenty of food for ourselves. We were also getting used to the idea of what to grow for sale, and how to sell it, although it was clear that selling piecemeal could take almost as much time as the growing; and as our profits were time-based, this would mean a near halving of already slender profit, plus the cost of petrol!

How much did we *need* to sell? Enough to pay our bills, obviously. Surely that would not be too hard to do? Even if, as I calculated once, we were sometimes working for rather less than 50p an hour.

It's not that easy, is it? For the sake of argument, let's say we need £1,000 to pay our basic, essential, bills and expenses. What crops do we grow? £1,000 worth of potatoes? Not really, because being organic we need to rotate our crops, so we can't grow spuds every year, and anyway, blight is endemic. What's more, we don't have enough land for such a low-value crop, and even if we did have, we couldn't lift thousands of plants by hand in the window available between ripeness and the autumn rains; and anyway, we didn't come here to become medieval agricultural slaves. We came to run a largely self-sufficient smallholding, growing good food for ourselves, with a modest surplus to sell to others at a fair price.

So what about growing three crops, rotatable? Spuds, cabbage and onions, say? Again, not enough land to grow enough 'cheap' crops like these on; and it's pointless to do expensive handwork on crops that can be mechanically planted and harvested. Nobody will pay the price you would be forced to ask, especially once a wholesaler and a supermarket have doubled or trebled it. And who in his right mind wants to spend his entire summer hoeing onions – ONLY onions - or picking caterpillars off acres of ONLY cabbages?

And then there's the question of *how* you sell your crop. Retail or wholesale? This depends on the quantity and the nature of the goods. Lettuces don't keep, for example. They have to be shifted immediately; and it's tough if another grower beats you to the shop or wholesaler first, because you don't get to sell the crop you've spent dozens of hours producing. So you need to work in conjunction with a marketer, somehow. Perhaps you grow what he knows he can sell; or agree that you'll supply a specific quantity of specific crops; or take him stuff on spec, knowing the deal will almost certainly be 'sale or return' … so who takes the risk, after all the time and cost of growing the stuff? You do. Hmm.

All very tricky. A neat summary of the marketing dilemma might go like this:

- Selling 10 lettuce: easy! Friends and neighbours will have them.
- Selling 100 lettuce: pretty easy. Friends, neighbours, pubs and restaurants will take them; but your delivery costs might be higher than your profit.
- Selling 1,000 lettuce: harder. Too many for local individuals and pubs. Not enough for a big wholesaler to want

to be bothered with (although you might strike lucky with a local man).

- Selling 10,000 lettuce: easy! They go straight to a wholesaler ... assuming you can pay to have them harvested, and own a lorry.

Obviously, these numbers are not meant to be taken too seriously, but they do help to show up the problem. Quantities and timing need careful thought.

\* \* \*

Enough of this boring economics!

In the spring of '84 Daisy did what Daisy ought, and produced her first pure Jersey calf for us. He was amazing. I've never seen anything so patently brand new. So shiny and glossy; big Bambi eyes; a nose like a sergeant's toe-cap; perfectly neat feet, all pale horn and gleaming.

Daisy was the proud Mum, although this was probably her eighth calf, and nothing new. She licked him and nuzzled him, and butted him gently round to her udder while he was too young and daft to know where it was (this period lasted a good ten minutes, once he was up and about – see below – after which he became a master of the art of docking to refuel). The kids fell in love with him and named him 'Wally' because, let's face it, 'Einstein' would not have been appropriate. Wally went on to thrive and become an ornament to the farm, at least for the nine months of his allotted span.

'Allotted span'? What's this? Who is playing God here?

Facts of life, brother; facts of life.

By autumn, he'd grown a lot and we didn't need Ken to tell us Wally was ready to 'go'. It was not an easy decision to take, however. He was still beautiful, if a little boisterous and not to be entirely trusted when your back was turned, as he might try to mount you to practise having sex on, or give you a testing little poke or two with his stubby horns, the way you or I might prod a melon for ripeness. Not as much fun as you might think, I might add, as he already weighed a quarter of a ton.

And we knew about grown-up Jersey bulls. They had one next door that glowered hate from every pore and bellowed flames deep into the night.

While Jersey cows remain pin-up girls throughout their lives, the bulls rapidly develop into real mean critters. Very heavy and aggressive mean critters. Meaner than other breeds, apparently. Real sling-you-over-the-hedge-with-big-stubby-horns mean.

So … Wally had to go. Nobody would want to employ him as a father. Only top pedigree jobs get to be studs and live a life of Ottoman luxe. Wally wasn't top notch.

We sniffed back our feelings when Ken arrived in his pick-up to take him away for us. 'Anyone want to come for the ride?' Sorry. Not this time. Maybe next time.

We tried very hard to be grown-up and rational about it, and very nearly succeeded. After all, wasn't he extraordinarily lucky to have had these nine months? Certainly, he was, because he'd arrived prematurely and taken us by surprise …

We'd rung Ken to say we thought our first calf was due, and described the symptoms 'Aye. No panic yet. I'll be over shortly.' We went for lunch.

Ken arrived in mid-sandwich. 'You got a little calf! Wanna come and see him?'

We all rushed out. It seems Ken had come across almost immediately and peeped into the byre en route. Wally was nose-down in the muck, and couldn't breathe. Ken yanked the bolt clean out of the fixing in his haste, and hauled Wally up and poked his nose with straws and shook him until he responded. Saved his life, no doubt.

But it wasn't over yet.

'He's a bit of a tiddler, isn't he? Don't seem convinced he ought to be here!'

'D'you think he'll live?'

'He might and he mightn't.'

'What d'you mean?'

'He ain't awake enough to get his grub. Ain't hardly awake at all.'

'So what do we do?'

'Well ... you can have some fun with him ... or you can let him *die*.'

That last phrase, couched in Ken's rich Radnorshire accent, said it all.

'Fun? What d'you mean?'

[CUE DRAMATIC CHORD, AS ALL PARTICIPANTS TURN TO FACE THE READER. CUT TO WALLY LOOKING TOO CUTE TO BE TRUE, BUT DANGEROUSLY DOZY ...]

The less patient reader might choose to skip a page or two to see what happened next, as the intervening article is all about boring vegetables, and has no strong element of Real Life

Drama. Will we have fun with Wally? What sort of fun?? Will Wally survive the night???*

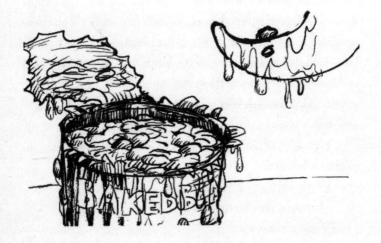

---

* 'If you've been paying any sort of attention so far, it is clear that he did survive. So you may wish to read the boring veggie article after all. Up to you.

# Commercial vs Garden Growing,
## Spring 1992

Growing organic vegetables commercially is a very different kettle of pickles from growing for the kitchen. A kitchen gardener selects crops for their variety, nutritional value and flavour. He also wants varieties which will mature gradually and sequentially and hold well 'on the field'.

A commercial grower has other considerations. Nobody can make a living selling a dozen lettuces here and a sack of carrots there, no matter how nutritious and tasty they are. In order to pay at all, vegetables have to be grown and sold wholesale, and this means growing crops and varieties that suit the wholesaler. If he can't sell Mongolian Glutinous Tripe-Grass then there's no point in us planting it. And when he wants lettuces, he wants thousands at a time, week after week. If he can't rely on the grower to come up with good, regular material, he doesn't want to know. And I can't find it in my heart to blame him; well, not entirely, anyway.

The only way to grow to this sort of specification is to use F1 hybrid seed, which will produce a uniform crop which will flush in $x$ days precisely, allowing for programmed sowing and cropping. We once asked Elsoms, a major commercial seed firm, for a cropping programme for protected lettuces to grow over winter in the polytunnels. By return of post they sent us a sheaf of charts and tables that would not have looked out of place in the Logistics Plan for D-day: hi-tec veg.

There is an obvious problem here. If strains are selected for uniformity and reliable cropping above all else, what is sacrificed on the way? All too often it is nutritional value and taste, and, by the way, genetic variety.

This is not to say that all F1's are valueless and tasteless, but the tendency is there: witness the pallid orange skittles tasting faintly of TCP and mothballs that are marketed as carrots by the chemical growers. Generations of people have grown to think that that is the way carrots *ought* to taste. On more than one occasion I've given a slice of one of our carrots to ladies of a certain age, and watched their faces light up in recognition: 'That's how carrots used to be when I was a kid! I'd clean forgotten!' Generations have grown up since with nothing to compare the current tasteless rubbish to. Now isn't that very sad?

Organic growing can maximise the potential of F1 and other soulless strains, but can only do so much. What can you expect to do with Moneymaker tomatoes, for example, which apparently were bred to have thick polythene skins to minimise transit damage? Or sprouts that have been so in-bred for machine picking that they have lost virtually all natural resistance to disease and need spraying more often than Dusty Springfield's hair-do? There is a case on record of some Dutch lettuces being condemned after being sprayed twenty-eight times. Was this a one-off? What do you think? And how often do you think the last lettuce / tomato / apple / *any*thing you bought has been sprayed?

It would be nice to see some serious work being done by the big breeders towards producing varieties that are high in nutritional value and taste, perhaps at the cost of absolute uniformity of shape. But as the cost of seed development and registration is so high, this is not going to happen until a lot more organically minded consumers insist that they want healthy food, not drug-addicted ornaments. And that won't happen until the public is persuaded to shop with its intelligence rather than with its eyeballs. How do *you* shop?

Growing organically costs a little more than growing chemically because of the man-hours put in, and the profit margin for the grower often lies in just those 'mis-shapes' which are assiduously rejected by the supermarkets. Thus, in order to show a profit, the price per unit of the 'perfect' specimens must be raised, to cover the wastage inherent in the 'mis-shapes'. And while organic produce on the supermarket shelf costs so much more than chemical goods, the shopper's intelligence will usually guide him or her into buying the cheaper, 'bargain', chemically-dependent goods.

In a nutshell, the supermarkets' addiction to cosmetic vegetables is what keeps the cost of organic veg so high that potential new buyers are consistently discouraged. Maybe new consumer group movements will help to destroy this foolish monopoly.

Meanwhile, small organic growers and allotment holders will be further discouraged from producing high quality local vegetables by the Soil Association's decision to raise the cost of the SA Symbol to £100 per annum. The SA has done a fine job in getting its excellent Standards accepted as the norm by the Government, but it seems to have been done at the expense of the small producer. Why is it not possible to devise a Symbol scheme for gardeners that everyone can afford? Has anybody tried? What has happened to 'Small is Beautiful'?

If any other HDRA members feel strongly about this issue, I'd be pleased to hear from them.*

Meanwhile, good growing! (April and the sheep send their regards.)

\* \* \*

---

* No response. Presumably no interest. Pity

See... told you it was boring.

But there are some quite important points in there, nevertheless. Ten years on, the supermarkets' monopoly is still effectively with us, though some of their sillier 'standards' seem to have improved, and organic veg seem to be proportionately less expensive than they were at time of writing. Small steps forward. But what a shame it is that we don't appear to have made any progress on the 'food miles' front.

It's a principle of organic growing, and part of the doctrine of the Soil Association – the guardian of organic standards – that food should be produced as close to the point of consumption as possible. It is a point of simple common sense: local stuff will be fresher, and will not have been involved in distributing cubic miles of diesel pollution while being shunted back and forth all over the country, or indeed, continent.

People are largely unaware of how widely travelled a humble carrot might be. A grower in Wales might take it 50 or 100 miles to his wholesaler; there it will be packed and then may go another 100 miles or more to a supermarket distribution centre in England; from there it might go another 100 miles or more to a retail outlet: possibly to the supermarket just round the corner from the farm where the carrot was grown. This bizarre procedure is normal for just about every veg in the country. And as for those french beans and pointless and tasteless baby sweetcorn which have been FLOWN in from Kenya and heaven help us, *Thailand* ...

I had a personal experience of this peculiarly crazy business one day when I was delivering our courgettes to the organic wholesaler. In the yard was the biggest articulated lorry allowed on the road, which was being divested of 30 tons of Spanish onions. I got chatting to the driver who told me that he was going straight back over the channel that afternoon, into Belgium, where he would pick up his cargo for taking back to Spain. 'What cargo will it be?' I asked.

You've guessed, haven't you? 'Thirty tons of onions, señor.' And he didn't seem to think there was anything even slightly ironic, not to mention totally lunatic, in this state of affairs. Most people share his view.

It's a great pity that we still need to import 80 per cent of our organic veg. Farmers whose livelihood has been threatened by the various recent pests and devastations seem unwilling to take the organic initiative without more government subsidies. The market is definitely there waiting for them: a neighbour of ours who did dare to take the leap told me his lamb sales alone brought in an *extra* £9,000 last year.

One glimmer of light is that some supermarkets are making positive noises about buying some organic veg direct from local growers. I hope they do so. One small step for sanity; one less 300-mile voyage for the humble carrot. And, of course, one dose less of fuel wasted. And one dose less of greenhouse gas.

Ten years on from when the original article was written, the minimum price for the SA symbol has risen from £117.50 (inc VAT) to £464.12 (ditto). I, as a smallholder, find this wildly excessive: it would represent some seven per cent of our projected maximum income. Any gardener would find it laughable or pitiful, I imagine.

Wouldn't it be nice if the Soil Association (perhaps in league with HDRA) could devise a way of accrediting bona fide gardeners and allotment-holders with the 'SA Organic' Symbol without charging them the same rate as that for a 25 acre farmer: £464.12 per annum? How can a genuinely ecologically-minded gardener legally produce 'organic' veg for sale in his local shop, while it will cost him £464.12 every year to do so? He might be assiduously sticking to SA principles, but unless he pays up £464.12 as an annual tithe, he is in breach of the law if he writes 'organic' on his box of runners. 'Small is Beautiful'? Not at these prices it isn't.

# Timeless ...

So ... did Wally survive his perinatal drama?

You will be relieved to know ... yes, he did. But it was quite a process.

Ken told us what to do. We built a little open-topped box of straw bales in one side of the byre, and scattered plenty of warm straw inside. Then I lifted little moribund Wally into his cosy bed, shiny new hooves and all, and covered him with Anne's old duffel coat. Then we lined up four empty one-gallon juice containers and Caity's old feeding bottle, and I prepared for a long vigil.

It was a brilliantly frosty night, and one I will never forget. Every hour I went out to check on him, and every two hours I replaced his rapidly cooling 'hot fruit-juice bottles' with two nice warm ones, and stroked and spoke softly to him. Daisy watched on, and occasionally added a baritone murmur. I'm sure she knew I was trying to help her baby.

It was warm and peaceful and alive in that old byre, with just the rustle of hoof in straw, softly lit by a single mellow oil lamp that cast the same shadows that inspired Caravaggio and Louis le Nain centuries before. And outside there was diamond-sharp moonlight on the frost.

Suddenly I was taken back a thousand ... two thousand years. This little drama had been enacted countless millions of times, all across the world and down the sweep of history. A sickly calf who needed human care. I felt a sudden affinity for every farmer and peasant who had ever lived. I shared a little in the life of the Masai; and in the life of the innkeeper on whose door Joseph knocked. And as I watched that beautiful sleeping little beast, I was also taken up with the amazing miracle of it all:

that a creature that spent its life chewing grass should produce not a heap of chewed grass, but this marvellous little being.

I think I began to really think for the first time in my life that night. Thank you, Wally, for that.

And yes, by morning he was beginning to stir. The warm sugary saline Ken recommended finally kicked in, although there were several moments of the purest slapstick, trying to make this totally bewildered little critter suck the liquid from a teated bottle. Up his nose, down his cheek, up his nose again, occasionally scoring a palpable hit amidst a flood of dribble. Then he learned, and sucked. 'How does he know he's supposed to learn?' I asked. And later, when he stood up, 'How does he know to butt upwards into the udder to start the milk flowing?' 'And how does Daisy, an utterly rigid vegetarian, know she ought to eat her own afterbirth?'

* * *

Nine months later, he tasted wonderful, and fed six of us and a couple of friends, for months. Thank you, Wally, for that, too.

# Organic vs Vegetarian? Summer 1992

April, daughter of Daisy, our doughty Jersey housecow, has just produced her sixth calf. As ever, she just had it. No problems, no vet, no panic. In fact, the whole thing was over in fifteen minutes, and the little heifer is a beauty. It still hurts, even after eight years of eating our young, to know that sooner or later she'll end up as meat.

'What!' I hear you cry. 'You heartless brute! How could you?'

Well, we were vegetarian for quite some time before we moved to the smallholding, but living and working here has driven us to the unexpected and apparently harsh conclusion that as far as we can see, you can either be organic, or vegetarian, but not both.

The logic behind this seemingly brutal claim runs like this: like all living things, plants need feeding; chemical farmers who produce the tasteless rubbish you buy at the supermarket, use chemicals; chemicals weaken the plants, kill off the micro soil life, and destroy the soil structure, so it eventually erodes away, just as it is now doing in East Anglia (Britain's own up-and-coming dustbowl) and across the world, at the alarming rate of millions of tons a month.*

So chemical farming is a long-term disaster. What is the alternative?

Organic farmers use compost, which feeds the soil that feeds the crops; so soil is built up, not destroyed.

---

* And once it's gone, it's gone for good. Now there's a thought. Never mind global warming; soil loss is humanity's greatest problem. And the less soil there is, the more they are going to blast it with chemicals, aren't they? Unless you and I can persuade them to stabilise their soil by going organic.

But here's the rub: it is virtually impossible to produce enough vegetable-only compost for the needs of a farm or market garden, and anyway, compost needs a nitrogen input to get it going. The nitrogen naturally comes from animal urine and dung.

There may be a way around needing this animal input, but there is no convincing evidence that I know of that it is really practically possible for an organic farmer to avoid the need for animal excreta. And anyway, it is clear that animals and plants were made for each other. The animal is designed to eat the plants, and then to fertilise the land it grazes on.

As a by-product of the essential nitrogenous waste, a cow also produces valuable milk, and, with a bit of extra work, butter and cheese. But … and here's the point … she will only produce milk after producing a calf.

Unfortunately, 50 per cent of calves are male. As one bull will serve about 40 females, what is going to happen to the other 39 bull calves? They can't be left to eat a whole herd's worth of grass, or the price of milk would double every year (work it out for yourself). What is more, 39 testosterone-stuffed bulls in one field would soon make the Battle of Kursk look like *Salad Days*.

On top of all this, the Artificial Insemination service has reduced the need for bulls even more. One super-stud can now serve hundreds or thousands of heifers, and without all that unseemly bellowing and drooling.

So … if you want organic vegetables, and I for one think you ought to on ecological grounds alone, you must accept animals going for early slaughter, unless you can find a logical flaw in my argument.

In practice, rather more than 50 per cent of calves have to

go off early. A milk cow will need replacing about every six or seven years as her productivity drops. A farmer will plan her replacement a year or two ahead, by having the cow (probably a Jersey or Friesian) inseminated by a dairy bull ... again, a Friesian or Jersey. For all her other productively lactating years, the inseminator will be a beef bull (Hereford or Charolais, say), and the resulting beefy calf, having done its job of getting the milk flowing, will go for meat after about twelve months, male and female alike.

It was a hard decision to accept, but we felt it was the only honest one we could come to. The only alternative to a mixed organic holding was a vegan one; after all, if you are a 'vegetarian' who still uses milk products, you are implicitly accepting the calving-milking cycle I've just outlined, and all its implications for the calves.

Veganism does not seem entirely natural, however. The physiological evidence all seems to point to man being an omnivore. I honour the ethical stance vegans take, but am not sure how veganism can fit into sustainable agriculture.

This has not been an easy piece to write, as I know it will offend a lot of people's sensibilities, but I think the point needs making and needs facing up to.

Other points to add to the equation are that our calves (and lambs) are individually cared for, and live peaceful lives with their mothers. They have, we believe, no ability to foresee the future. Hence, they live happily until it is time to go. Here, though, there is a definite problem. Ideally, the calf should be separated, stunned, killed and butchered on the farm. Unfortunately, this is not possible, according to the laws of the land. Instead, it must be driven to a distant abattoir, which is less than satisfactory for rather too many reasons.

If people stopped using animal fertilisers, there would be other consequences too. Firstly, there would be an even greater use of ecologically damaging chemicals. And secondly, the countryside would alter radically. There would be no soothing fields of grazing cows or sheep. In fact the cow and the sheep would instantly become endangered species. And pasture land, with its rich flora, would soon revert to brambly scrub … or of course, turn into golf courses for the newly rich millionaire executives of the Agrochemical conglomerates, and Jolly Theme Parks for their families to play in.

Do I protest too much? Possibly. Yes, it is a sad day for us when the calf is trundled away up the drive; but all of us, farm animals, pets, and people, have to make the journey one day. And isn't quality of life more important than mere quantity? It is for me, anyway. Who wants to 'live' for a hundred and fifty years in a sterile cabinet, actual or metaphorical?

Just to complicate matters, there seems to be quite a lot of evidence that even plants feel something akin to shock and pain. Certainly they can be chloroformed for easier transplanting.

Where does one draw the line? Is there a line to be drawn?

And upon that philosophical note, I'd better get on with fetching April her breakfast. Hay, today. She'll like that.

* * *

I've thought over the Organic vs Vegetarian issue quite often since writing that, and still come to very similar conclusions. If you want sustainable fertility, it seems there still is no real alternative to keeping animals, with all that that implies.

But what mysteries soil and growth are. Our local substrate is a schisty-slatey sort of stuff. New roads hack cuttings through it, and expose the virgin rock, revealing planes of possible fracture within it; but it's still solid rock. Within a few years, however, there are six foot high trees growing out of it.

Where is the carefully balanced NPK (the nitrogen, phosphorus and potassium fix so beloved by chemical farmers) coming from? The trace elements? The billions of micro-organisms we are told a fertile soil needs? All these trees get is light, air, (plenty of) water and whatever it is they can glean from solid rock and the tiny fissures in it that their roots grab into. Are trees different from other plants? Do they need fewer nutrients and symbiotic bugs? I'm very impressed, whatever.

# My kingdom for a ... !

Our first calf was just the start of the new directions 1984 brought.

We'd done a lot of thinking about mechanisation and how we really did need to have some basic machinery. The rotavator was very good for smallscale work like cultivating a seedbed, but it was inappropriate for turning over a whole field. Even we knew that. We really would have to get a tractor.

But what sort? In fact we already had a tractor that we bought on site with the house. It was a venerable Fordson Major, with a Quicke loading bucket on the front. A useful bit of kit for someone with tons of muck to shift out of a big shed, but no use at all to us. It was just too big.

Our ideal would have been one of those little Japanese Kubotas that were just beginning to appear: something that would plough and harrow an acre without needing another acre to turn round in, and which could haul a modest trailer, so we could take the kids and Wally to the seaside for the day. But we quickly discovered that 'smaller' does not necessarily mean 'cheaper'. We couldn't afford one.

We talked to The Oracle: Ken. Like most farmers, he had two tractors: one for big work and one for pootling about in. The big Zetor was even bigger than our Fordson, but the little Massey-Ferguson looked promising. I can't remember the ins and outs of the discussions now, but it ended with Ken convincing us that what we really wanted was a legendary Little Grey Fergie.

Nothing's easy, is it? Did we want the petrol-engined version? The diesel? Or the mythological-sounding 'Petrol/TVO hybrid'?

'TVO' is Tractor Vapourising Oil which nobody has made for 50 years.

'Don't worry,' said Ken. 'Heating oil; kerosene; same stuff; cheaper'n petrol; works just as well.'

Diesels have always been the most popular type as they run on tax-free 'red' diesel, and are mechanically less temperamental. Thus they are harder to find for sale, and when you do find one, it is likely to have had a harder life than the Flying Dutchman and be unfit for further service.

So, to our surprise, the TVO model seemed the best for us as it was cheaper to run than the petrol-alone version. We scoured the local papers, and almost immediately found a likely candidate.

Dad and I went to see it. It started easily, and the man showed me how to switch over from petrol to TVO when the motor was warm enough, then how to turn back to petrol before you switched off, so it would restart when cold next morning.

The clutch was fine, and nothing rattled excessively. The power-take-off shaft (pto) went round and round, and the bodywork had a substantial and stylish anti-roll bar that I could mount a Lewis gun on to chase off the RAF Harriers that strafed us with megabels from time to time. It also had a useful 'link-box': the equivalent of a modest saddle-bag on a bicycle, that could be raised and lowered on the hydraulics. Brilliant. The only minor drawback was that it had been repainted; not in the traditional Fergie Mid-Grey, but something a little livelier. 'Eggshell?' 'Pastel Cerulean?' Alun thought 'Pooftah Blue' was about right. Never mind. *We* liked it.

So ... back to the point of sale: all essentials in working order, as established by my rigorous three-point (does it start?

does it stop? is anything obviously missing?) checklist. And at £250 it was a bargain. *And* the man threw in a knackered old string vest of a chain harrow which we slung, very heavily and slowly, into the link-box, as casually as to the manner born.

I drove it the ten miles home, as happy as Larry in the bright spring sunshine. You can keep your sports cars. Say 'open top', say 'tractor'; say also 'coat buttoned up to the neck' and 'substantial hat, knotted securely under the chin'.

And anyway, they say the Fergie engine is the same as the Triumph TR3. Vroom!*

I discovered that it had no brakes worth mentioning as I came down our drive. An alarming moment, as the right-hand brake was even less worth mentioning than the left-hand one, and I began an interestingly urgent slew towards the ditch, tears, and expensive repairs.

'Oh, husband dear! What have you done?'

'Crashed the adjectival tractor, haven't I? Why don't you go and knit a sock, or something?'

Luckily, I was going gingerly, and managed to straighten the steering and slow my progress enough by crashing the box into a lower gear before the slope got too steep. No syncromesh, you understand. I crashed the gears for several years afterwards, actually, until someone told me you don't change gear on a tractor. You start in the gear you need for the task. Simple. No wonder I sometimes got odd looks, but I bet I could still beat the biggest Zetor in the country away from the lights.

Ken told us what to look for in a second-hand plough, and again, we found a beauty within a week for just £25. I had

* Different gearbox, I suspect, or I could plough an acre in four minutes flat, especially if I could master the hand-brake turns.

no idea how to set it up or adjust it (still haven't) so Ken spannered and hammered things until they suited our particular circumstance. Ten minutes later I was learning to plough.

I think ploughing must be the agricultural equivalent of plastering a wall. They are both jobs that require great subtlety of skill, but which leave very little to show for it at the end, just a sort of bland tabula rasa. Our top field curves down at an irregularly declining rate, which means that you need to shift the cutting-depth lever imperceptibly as you plough from the top of the field to the bottom. If you don't depress it enough, the blades come clean out of the ground, and you have to reverse and re-engage, which can be a messy and unaesthetic business. If, on the other hand, you're too heavy-handed with the lever, the shares dig too deep and your big rear wheels skid and spin, and the whole machine starts bucking and bouncing, and everybody laughs and points scornful fingers. Har Har; you try it, then.

Modern tractors have automatic sensors to keep the depth constant; also full quadraphonic sound; satellite phones; auto this; auto that; a deep fryer for chips and doughnuts; and a cab like a Concorde cockpit.

Working with the Fergie was more like flying a Sopwith Camel: seat of the pants stuff. I eventually learned to adjust the depth lever according to the revs of the engine and the steadiness of the pull; I needed to pay full attention, rather like I imagine the old ploughman needed to, following his horse. Proper ploughing. Very rewarding, too, when/if you finally get it right.

The other good news for early 1984 was that several local organic types had ganged together to form a producers' co-operative. Things were looking up …

# Water Problems, Autumn 1992

Water, water, everywhere …

I would never have believed how important water would become in our lives. We arrived here fresh from the suburbs where water was a forgotten commodity: it was ad lib, on tap, and paid for by a quarterly set bill.

Down on the farm, things are different. We have both metered mains water and free spring water. A great improvement, you say? Yes of course, but there are snags. For example, the mains meter is half a mile away, across two other people's fields, and the pipe is at constant risk of being damaged without our knowledge, but at our expense.

Obviously, we use spring water most (an analyst once said it was the purest water he'd ever come across) and have even considered bottling it for sale as 'Eau Gorbleimu!' or 'Eau Calcwtta' or somesuch, with a suitably decorative and eye-catching label.

It's only a gentle trickle, but it rarely runs completely dry. It does rise in another party's field, however, and as he uses chemical fertilisers occasionally, we turn to mains for a few days when we see him out there, spinning his NPK.*

Once in a while the spring supply inexplicably seizes up for a couple of days, until, with a spurt and a shudder, a tiny worm or part of a froglet shoots out of the tap and we're back to normal. Ah! Pure water!

In winter we have to leave the cold tap in the bathroom trickling overnight, as the spring supply pipe lies on the surface for a hundred yards, and freezes solid as soon as the

---

* NPK: Nitrogen (for leaf), Phosphorus (for root), Potassium (for fruit). The Trinity from which all chemical farming flows.

laws of physics decently allow. If this happens, we have to switch to mains.

This can cause problems, as the ballcock valve in the cold tank is not strong enough to cut off the mains pressure completely, and the excess steadily drips into the overflow system. On a cold night the overflow pipe freezes up, the water backs up the pipe, and eventually spills onto the bedroom floor. Would you believe it?

Change the valve? I seriously think it is only accessible through the roof, probably with the aid of scaffolding. I'm pretty sure the man who installed that tank was the same genius who decided not to bury the spring supply pipe; and who thought he could divert a stream and replace it with a drive; and who ran the central heating pipes under a concrete floor to the bathroom, then *outside* the house to the kitchen. I would like to meet him one day. On second thoughts, perhaps not. One thing the world does not need is another ghastly murder, with evidence of extended sarcasm and gross verbal abuse.

The pressure differential can be a nuisance outside as well. We have gradually built up a pretty complex network of hosepipes and taps, delivering water to all three polytunnels, the animal watering troughs, and the veg patch. The Hozelock-type connections work fine on the spring water, but they blow apart like trip fuses if we switch to mains, flooding away hundreds of gallons an hour if we don't spot the problem in time. So we stick to spring water wherever possible.

The gentle pressure is excellent for the leaky hose irrigation we favour in the tunnels, but not quite strong enough to allow us to irrigate the top of our sloping veg patch. We only really need water for planting out the courgettes, however, so this is not much of a problem. Normally we don't water any of our crops once they're in. Even in the driest summer they get their roots well down and never seem to suffer. I suspect that the field is on a minor spring line. Our drive and yard certainly are. Damp patches and holes are always appearing in them, occasionally glazed with a film of attractively iridescent weed, and smelling of the seaside.

I once made a list of all our water-based problems, and stopped at thirty-two, including three leaking roofs in the house alone (all since solved, mercifully); eight sources of wet in the sheds (usually faulty roofs, but also including seepage, rising springs and a ditch overflowing; again, mainly solved); and, most bizarrely, a mysterious wet patch on the sittingroom carpet, for which poor Porky got the initial blame, but which was eventually traced to water trickling down the inside of the television aerial cable. Could you make it up?

Since then, apart from myriad minor hitches, we've had the bathroom flooded twice (once from above, once from below), the hot tank threaten to explode in the middle of a freezing night, the ditches by the drive dredged of six tons of muck, yards of the drive's tarmac veneer washed away, etc.

And, oddly, within a six-week period, our car needed a new water pump, a new thermostat, and a new cylinder head, which had been corroded by … water.

For the benefit of any astrologers out there, Anne and I are both Cancerians; a water sign, I believe. I used to be sceptical about astrology.

But it's not all bad news; the spring water makes excellent beer.

* * *

… and the first faults our next car developed were: a leaking header tank, then cylinder head corrosion, again caused by … water. 'Never seen it in a Volvo before,' said Alun the Mechanic. We had it twice, within weeks.

Water continues to be an occasional preoccupation. As I write these words there's a builder trying to find an airlock in the central heating system. How did it get there? Anne was cleaning up the utility room, shifted the washing machine to reach behind it, and knocked an ancient and redundant and long-forgotten stop-cock at near floor level. It immediately started dripping. But of course; what else! The builder stopped the drip, but now we've got an airlock, and it's going to take all day and cost a lot of inflated plumber-money to fix. Hey ho.

It just had to be a *washing* machine, didn't it?

# Water, water ...

Dwlalu Farm used to be considerably larger. Right up till we purchased it, it had a good 40 acres. The fields on both sides of the drive belonged to it, and so did the two big fields between the top of the drive and the main road. Thus, the mains water pipe ran from the meter, which is dug in at the side of the main road, under the adjacent field and down to the farmhouse, running under the fields bordering the left side of the drive as it went. All on Dwlalu land. Fine.

However, the owner we bought off had been advised to sell the house with just a couple of fields, and dispose of the other six fields separately. This, we think, is because they wanted a quick sale, and purchasers for 40 acre farms were hard to find at the time.

This was good for us, because it meant the expensive and unaffordable farm had become a cheaper and affordable smallholding. However, the sale of the other fields meant that we no longer had control over our own mains water supply. 'How much does this matter?' I hear you ask.

Well, for example, it was several months before the penny dropped that our neighbour on the left had been watering his gaggle of pedigree Jersey heifers, also for several months, at our expense. Oh dear. How do we handle this one? Confrontation and demand for payment? Send the sheep round to stare at him? Secretly dynamite his slurry pit?

In the end I just walked over and talked to him. He was quite apologetic. Said the issue hadn't crossed his mind, and paid us an approximate refund. The following week, we agreed that it would be better if he disconnected his field baths from our mains water supply completely, and ran his own pipe out to them. This was not down to any unnatural meanness on our part. It was to do with having control over our expensively metered bills. What if a bouncy young heifer kicked the pipe off the bath one evening, just for fun, and it went unnoticed for a couple of days? The wastage would show up like a mountain stream on our bill. Our budget was tight enough without that sort of extra worry.

Then those neighbours left and new people moved in. One morning we noticed that the bath in the top field had gone and that all the adjacent area had been mechanically disturbed, exposing the tied-off spur of our mains pipe that had once fed the bath.

The favoured means of temporarily (and often permanently) closing off a 'live' pipe is to bend a foot of it back on itself and tie it tight with baler twine. But such a closure is very vulnerable if it's just sticking up in a field. Doubly vulnerable in someone else's field, particularly if they don't know what it is, whose it is, and why it's there.

Dad decided we needed to make a more permanent closure and then safely bury the pipe, out of harm's way. OK. Let's do it.

We drive 200 yards up the drive, examine the site and make our plan.

Before we bury the pipe, we need to check that there's been no damage already done. Is there a leak anywhere? We should turn everything off, and check that the meter has stopped registering a flow.

So we drive the 200 yards back home and Dad makes sure he's turned off all his appliances. Then we drive the half mile to the mains meter. This lives in a little concrete bunker, sunk into the overgrown verge by the side of the main road.

We find the cover of the bunker, but also find we have nothing with us suitable for lifting it with. So we drive the half mile home, and collect a crowbar, and also a torch. A torch! Good thinking. Then we drive the half mile back to the meter.

Off comes the cover. The meter is right at the bottom of the bunker, not surprisingly perhaps, and you can only read it if you lie flat on your face in the foot-long herbage and peer intently. This means your feet are sticking out into the main road, so Dad acts as flag-man to warn traffic off.

I wonder what passing drivers made of it? It must have looked like a weird traffic accident. Or possibly someone being terminally sick into a convenient ditch, after eating a suspect pie for breakfast. It's happened before.

The torch is predictably on its last legs, but by the little pool of brown light it casts, I can see that the hole is half full of earth, and swarming with ants. How all that soil could have got in there, I've no idea. I try asking the ants, but ... well, you know ants. They love orders; hate questions.

The meter is completely buried. Do we have a trowel? No. We have a spade, though, but it does not fit the hole. Are we going to drive half a mile each way again, just to get a trowel? No, not if we can help it. Dad finds a plastic cup in the glove compartment which serves. Just.

So I bail out 5lb of anty soil (which unfortunately does not obey the laws of anty-gravity), rather carefully, as the ants look pretty vexed and grumpy, and aren't best pleased at having been questioned earlier. They enjoy going up my sleeve, however, which leads to me leaping up and down and swinging my arms round and round like The Semaphore Man Who Drank the Local Brew. More care is required and applied.

There is the meter. It is moving. Water is flowing. Oh dear. So we drive the half mile back and check again. Is something turned on and running after all?

Nothing is on in Dad's bungalow. Where then? It must be a leak, because we don't use mains in the Big House. BUT! as it turns out, the spring pressure has fallen so low that Anne has just this minute turned over to mains to water the tunnels.

Off.

Half a mile drive back. Peer into hole. Torch light shrunk from sienna to burnt umber. Meter has stopped. Excellent. We give it a few minutes. Still stopped. No leak. So we turn the stopcock off and drive the 600 yards back to the vulnerable pipe and screw on the big tough plastic stop-end thing. Then we bury the whole gubbins as deep as we can, with the help of the spade and plastic cup.

Then we drive the 600 yards back to the meter hole and turn the water back on again; then drive the half mile back home, well pleased with our work.

Five-minute job, as predicted.

So both of those baths are, we hope, now permanently sealed off. We just hope another new owner doesn't have any over-ambitious plans for moling or sub-soiling or mining in that top field. If he does, he'll strip out our poor flipping pipe like a varicose vein.

* * *

The tractor made a huge difference to our ability to prepare the land in the spring. My ploughing went through all the usual stages you'd expect, from wavy and dippy, through patchy and cloddy, to adequate and workable. The worst part to plough was just towards the bottom of the field, where the curve subtly alters. It took several years to crack that one, partly due to a lack of natural talent, but also because I only really had one day's practice a year.

The first time I did it, I managed to sort of flay the turf and sod clean off the subsoil. It just lay there on the surface, inverted, like a tired old anaconda on a pudding-stone beach. I tried going over it again, but that only made matters worse by crumpling it up into a weedy heap. I'm glad Ken didn't see it before I'd hauled it back into place and straightened it out. Terribly hard work, all by hand: bits of sod dropping off all over the place, and the resultant lines were anything but straight, but what was the alternative?

The eventual solution, partial at least, was harrowing. I'd never previously had any proper understanding of what harrowing was, or what it was for. I was about to find out.

# Muck and Spreading, Winter 1992

'Five acres and a cow' was once considered to be the minimum requirement for the peasant farmer, and it's true to say that any modern organic smallholding still revolves round the humble cow.* And I must mention that one or two of them are not so humble; our Daisy, for example, is definitely her own cow. If she decides to take four or five huge gobfuls of prime garlic seed cloves out of a carelessly accessible tray … you won't stop her. And you won't want to drink the milk for the next three days, either.

A lot of smallholders keep goats, arguing that they give quite enough milk for a lot less trouble. True, but we are not just talking dairy produce here; we are also talking serious dung. Despite their heroic efforts and questionable motivation, goats just ain't up to the job.

---

* ' …or goat.' Yes, I know. But as I patiently and rationally explained earlier, goats are clearly the spawn of the devil, etc.

An acre of veg can use ten tons of rotted muck a year, or twenty tons if you can get it. It's the best fertiliser of all, but it does bring its own problems, mainly involving distribution. You simply can't shift ten tons of muck uphill with a fork and barrow. Well you can, of course, but it will take up days of valuable sowing time, and it's bound to rain before you've finished.

You really do need mechanical help. Our amiable 1952 Fergie is our prime mover and a clapped-out old Massey muck-spreader is our precision instrument.

The cow and calf overwinter on straw with occasional top dressings of shredded Daily Telegraphs, contributed by Dad. This bedding, which is several inches thick by spring, is forked into the muck-spreader ('McSpreader', for readers north of the border) and sprayed out against a wall to aerate it, before being covered with black plastic and left to rot.

Meanwhile, a second heap, which has already rotted down and is now filling up with brilliant jiggling brandling worms, is forked onto the muck-spreader, along with 20–30lb of ground limestone per load, and sprayed or splattered onto the field, to give a rough pebble-dash effect. Very, very attractive.

The sheep and cows have already grazed off any greenery, and if we catch the weather just right, the top dressing of composted muck gets ploughed in, just four or five inches deep. That's where all the action is. No point in going any deeper.

That's in an ideal world.

In reality, the tractor won't start. It's used so rarely that it gets stiff and morbid out in its shed and doesn't much fancy any exercise come the spring. I run through my Tractor

Starting Ritual: tweak the plugs, emery the rotor arm, charge the battery, prime the carb with fresh petrol; and eventually Fergie bursts into joyous and throbbing life.*

Anyway, once the tractor is going, we can see about welding up the muck-spreader. Every year something else falls off, yet somehow it keeps going.

We fuel up the tractor with petrol in one tank and kerosene in the other and fetch the compressor to pump up the tyres. Then we reverse into the bottom yard and load up with delicious-looking compost, larded generously with buckets of ground-up limestone. THEN ... we roar off onto the field ... and it's a 50:50 chance we get stuck in the mud at least once. The other constant annual panic is that I will forget to fill the radiator after draining it over winter.

The actual job of muck-spreading takes no time at all. Five minutes per load; say thirty-five minutes to do the whole field. But in order to get simple old-fashioned rotted muck onto a small field we need a tractor, spreader, electricity supply, welder, compressor, black plastic, oil drums full of kerosene that you can only buy by the hundred gallons, and a mechanical genius cum Merlin living round the corner.

In a word: if we want to use Green fertiliser, we have to use quite a lot of un-Green equipment. Is this a paradox? Or is it simply a matter of using machinery appropriately? There's plenty of scope for philosophy out in the sticks.

---

* *Almost* every time. But sometimes sloth and despondency triumph over the mere diktats of physics and we need someone to point the bone. Last year, Alun the Mechanic drove down in answer to my desperate plea. He pointed a finger at the tractor: 'That's enough of that. Now start.' And it did. He didn't charge a fee. How could he?

# The joy of spikes ...

Once the muck and lime are spread, the muck-spreader has done its task for the year, and goes back into its tin hut for another 364½ days.*

Then it takes the rest of the day to plough the acre or so. Usually this stretches over two days, because, well ... it involves machinery, doesn't it? 'Time-saving' machinery, which is to entropy as babies are to vomit and stickiness: utterly inseparable. But, to be fair, the Fergie is remarkably reliable given that it's now 40 years old. How old's your car?

Once the plough is unhitched, the real fun can begin ... harrowing!

In essence, the plough inverts the soil and sod, and the harrow flattens it back: grass underneath, soil on top. Thus the grass becomes green manure, and the overturned soil is ready for being crunched into a tilthy seedbed by the rotavator.

As you might imagine, all these procedures call for some fine judgements of depth, which take time to learn in a Sopwith Camel.

'OK Algy! Give her a swing! Just off to have a harrowing experience in the top field. Port and lemon waiting? There's a good chap.'

There are quite a lot of different styles of harrow. What they all have in common is spikey things. A gorse bush dragged behind a goat would count as a harrow, at least until the goat ate it; it is of the same essence as the huge power

---

* At least it used to, until we found the tin hut collapsed under the weight of snow one sunny April morn. Fortunately I'd left the spreader out that night. The tractor was still inside the shed though. Ugh.

harrows that vibrate and tingle and shatter their way through endless miles of prairie. Just spikes.

The chain harrow is a lightweight: think 'chain mail for Bernard Manning'. It is really at its best spreading cowpats or molehills out over a pasture, or preparing a moderate seedbed on suitably level and fluffy soil. It is not at its best being hauled back and forth over bad virgin ploughing, especially when most of its spikes are worn down to nub-ends. A toothless dog gives a very poor bite, as every milkman or cano-masochist knows.

All I really achieved with the chain harrow was the unhelpful act of compressing the soil flat with the weight of the tractor, and capping it with a hard crust. Where once we had a field, we now had a carpark. The harrow's only contribution was to snag on any big sods immediately visible (bad ploughing leaves quite a few, it must be said) and drag them round the field. Once they'd dried out in the wind, it looked like some sort of dreadful hedgehog massacre. ('Suffer not an *hedgehog* to live, lest it devour multitudes of mollusks and stomachfooted creatures to the detriment of the multiplication thereof.' Deut 6:12.)

We found other odd bits of inter-locking spike harrow in various hedges and ditches and wired them together. Proper serious spikes, this time, despite the rust. In themselves they were not heavy enough to cut into the soil as much as we needed so we forced them down a bit by tying breezeblocks on top, which certainly helped, but not quite enough.

Consequently, just the once, Anne had a brief but bone-jarring ride standing on the breezeblocked harrow, Boudicca-style, with hornèd helm and reins of plaited baler twine.

Fortunately for her, her weight was, in this instance,* excessive, causing the spikes to dig too far in almost immediately, thus causing the high-revving tractor to 'come bulling', which means rearing and bucking alarmingly back onto its rear wheels, waving its front wheels helplessly round in the air for a while before crashing back to earth as the driver panics and switches off everything he can grab at or poke.

I'm not sure how Anne regarded the immediate prospect of being flattened under an inverted tractor, as she tends not to discuss it.

### Hippy and Wife Crushed and Mangled Horribly in Bizarre Tractor Death Scenario: Suicide Pact and Weird Ritual Suspected
### See Special Colour Supplement for Sketches, Diagrams and Forensic Enlargements

The spike harrow was definitely an improvement on the clapped-out chain version, but still rather too rough and ready to cope with our stony soil and my gung-ho life-on-the-ocean-wave approach to ploughing.

Then we heard of the spring-tine harrow; nay, we discovered that that heap of rusting scrap by the gate actually *was* one.

Broadly speaking, there are two kinds of tractor implements; those which are simply dragged along, and those which are mounted on the rear hydraulics which are used to raise and lower the implement both for transport and for reasons of depth control when operating, or turning at a head-land.

---

*And *only* in this instance, I feel I ought to point out quite clearly.

Our chain and spike harrows are pre-war devices, whose design is unchanged since pre-Roman times, that just drag along behind the slave, or ox, or tractor.

The spring-tine also drags, but it is a cleverly designed tool that runs on skids. The harrows (tines) are slightly curved, and are attached to curved springs mounted on rotatable bars. Thus, at the swing of a lever, the depth of harrowing can be (moderately) controlled. The springing helps prevent the tines digging in too far.

To my surprise, the first farmers' shop I tried still stocked spare tines. With a bit of ad hoc welding work by Alun, we had a mid-tec tool.

It was just about right. As my ploughing improved, the adjustable spring-tines really came into their own, and fluffed and lifted the soil as harrows are meant to. The only snag was that they also dragged up and collected stones. What should we do with them? The topsoil was thin enough already. If we took the stones off, it would be thinner still.

Did we need to take them off? Were they causing any real problem? We decided not. In fact we decided they were positively useful as they suppressed weeds, retained moisture, warmed the soil, and provided spacious accommodation for nice beetles. There's nothing like making a virtue out of a necessity, is there?

So ... properly tooled up for cultivation at last.

I became pretty adept at harrowing and thoroughly enjoyed blazing round the field in top gear, in my Red Baron helmet and goggles, white silk scarf streaming behind me, thoroughly locked into my other persona of Roger ('Rover') N. Dowt, the Ace of Abbeville, performing ever more intricate arabesques and sworls, scattering desiccated hedgehogs everywhere, and Lewis-gunning any Harrier or A-10 Tankbuster that dared show its tailpipe.

Only once did I misgauge my course.

At that precise moment I was Michael Schumacher, approaching a particularly nasty chicane at Monaco. Cut revs; hard a-port; …oh dear. Not going to make it. Oh dear, oh dear.

I didn't actually *hit* the fence, but snow-ploughed to a dusty halt within four inches of it. No problem. Slight embarrassment, that's all.

The snag was that the harrow was of an old design, and was thus not liftable on the hydraulics. Thus I couldn't reverse with it on. Well I *could*, but only by driving over it, thus crushing the harrow and puncturing both tyres, which scarcely seemed worth the effort. So I had to face the ignominy of disconnecting the harrow and hauling it slowly out of harm's way while I rescued the tractor. Daisy leered at me over the fence, and Wally wanted to join in, but I wouldn't let him.

### Hippy Can't Even Drive a Tractor!
### Nearly Smashes Own Fence to Pieces!
### Artist's Impression, p8-9

# Orchard, Spring 1993

One of our top priorities when we first arrived here was planting an orchard. We consulted every book we could find and drew up a list of likely candidates, checking for diploids and triploids, and flowering periods and making sure that the early bloomers were upwind of the late bloomers. We planted sixteen apples in all, on semi-dwarfing stocks, including two scions from an unnamed parent we had grown on for us by a specialist. We had fond memories of this apple from our old garden. Huge, pale green orbs, amazingly juicy, but likely to go yellow and tasteless after a week or two.*

We followed Lawrence Hills' (of HDRA) advice and planted them all on to a base of subsoil well mixed with broken shell, hoof and horn, bonemeal, and handfuls of hair donated by the local barber (and, I assume, his customers). We staked them firmly to the leeward of 52 inch fence posts, well ligatured with old tights, and moved on to the other trees: four pears, five plums, two peaches, two cherries, a mulberry and a quince.

Ten years later, the orchard has changed from a place of great peace and optimism into a place of pleasure, but also some frustration and occasional angst.

The peaches, the mulberry, three of the plums and two apples didn't make it.

A replacement peach also died; so did an apparently flourishing apricot.

There's nothing wrong with the soil that we know of. Grass grows wonderfully. Perhaps it's the spring/autumn/

---

*'We never did find out what variety they were. 'Dolly Parton' was suggested.

winter winds? Or is it just that we were too optimistic, hoping for fruit at 550ft?

Well, I don't know. We certainly have had fruit, particularly apples, and particularly from the anonymous orbs, but it's pretty hit and miss overall.

The pear trees refuse to grow beyond teenage size, while the Oullins Gage grows to 15ft but produces only a handful of fruit. Both the other gages died for no clear reason; the Czar fruited mildly for a year or two, then packed it in and retreated to the Astral; but Victoria usually does pretty well. I see no clear pattern.

Again: the Morello cherry never quite dies, but never quite does anything else either, while Stella reliably produces a small amount of beautiful fruit almost every time. The quince plods gnomically on; windswept, crouched and terribly Japanese-looking.

Even more oddly, for the past two years the apples have set fruit, but when we came to pick them, they'd all disappeared. Yes … we wondered who'd pinched them, as well. But I don't think anyone could have. Squirrels? Magpies? Wallabies? Daytrippers from the planet Zog?

The trees have had more than a little aggravation over the years. First they were all debarked at least somewhat by rabbits, till we made them chicken-wire collars (the trees, obviously). Then the cow got into the orchard and did some freestyle pruning. The sheep have added their own twopennorth as well, despite us encircling* the trees with old pallets. One or two sheepy geniuses reach up on their back legs to reach the foliage.

---

* 'Ensquaring' would be more accurate; 'entriangling' more accurate still.

Why let the sheep in? Well, we have to keep the grass down; and the best way of doing that is to use it creatively. I did once Flymo the orchard, but I've tried to keep quiet about it ever since. What came over me? Grass is food, not a job.

We did keep geese in the orchard until a fox got them one Christmas. So, faute de mieux, sheep it is.

Although the trees have had their share of bad experience, they did get off to a flying start, and should by now be coping nicely, if not wonderfully. But there's a lot of canker, a lot of infestation by small caterpillars and weevils and a lot of, well, I don't know … lack of commitment. The trees look pretty good, especially at pruning time, but somehow they're not putting their hearts into it. Should we prune them this year? I'm not sure. Perhaps they need a little more creative neglect. If any tree-fancying HDRA members find themselves in our locality I'd be grateful to hear what advice they can offer.

Meanwhile, perhaps we should consider specialising in hunchbacked quinces.

*Editor's Note: I feel sorry for Chas, but I do think he was a touch optimistic expecting pears and peaches to grow half way up a Welsh mountain!*

\* \* \*

Point taken, Editor. But we were young and gay, and the world was our lobster, and we were ready to try new things and challenge convention.

Perhaps pears and plums hadn't grown there before because nobody had ever actually tried? Perhaps the varieties we chose would be the key? Perhaps the microclimate might be unexpectedly favourable? And perhaps our pigs, when we eventually got round to it, might well be taught to fly? If not to actually loop the loop, perhaps to at least glide gracefully to a feather-light four-point landing from a running leap off the top of the black barn? Optimism was in our heart.

One kind soul did take the trouble to offer his help after reading the article. He had a look round, suggested that our best bet was to clear the grass from a 3ft circle round the base of every tree before manuring them well, and left, without taking up my offer of a cup of tea. I don't think he was terribly impressed with our efforts so far.

We were just too busy to clear all that grass, but we did mulch half a dozen trees with sections of thick rubber matting to see if it made a difference. It didn't. And we couldn't spare all that manure, either, particularly as we knew from experience that any manure laid round trees would instantly be scattered the length and breadth of the county by scrabbling free-range chickens.

The trees would have to make their own way. I think we'd already begun to feel that our efforts were unlikely to improve matters very much.

To begin with, we'd pruned and trained the apples assiduously, but they didn't seem to gain much from it. Certainly, they didn't become any less cankered. The books say to cut canker out or lop the branch. We couldn't do that or we'd have no trees left at all.

'And that's the Cox's Pippin …'

'Where?'

'You're standing on it.'

'Oh, sorry.'

# A fruity bit ...

We took an awful lot of trouble choosing those varieties and learned rather more than we needed to know about the sexual habits of Bramleys and Russets, and the curious world of the 'triploids'.*

Just for the record, the varieties we settled on were: Katja, Sturmer, Adams Pearmain, James Grieve, Claygate, Cox Orange Pippin, Bramley, Crispin, Egremont, Ribston, Grenadier, Court Pendu Plat, and Crawley Beauty. We also had the two graftings of the Juicy Green Orb jobs we'd brought with us.

Our theory was pretty good, we think. It just didn't work out quite so well in practice, rather like the Weimar republic. That didn't improve from pruning and cosseting, either.

Twenty years on from the original plantings, all the apples bar two are still plodding on; the fatalities were the Juicy Green Orbs. Quite why they and only they succumbed, we may never know. Otherwise we still get a regular crop of decent fruit from those poor broken-winded canker-ridden trees. The only ones that are a bit over-feeble are the J Grieve, the Crawley and the CPP, which has never produced a decent-sized fruit in its life, as far as I can remember. Mark you, it's on a particularly bony crest of land and perhaps, being Elizabethan or older, it produces mignon fruit by nature.

The pears we picked (rather too rarely, as it turns out) were Conference, Doyenne, Williams, and Josephine. They are still alive, but not inclined to get involved. They give us a

---

* Didn't they have a minor chart hit in 1982 with *A Pippin for the Teacher*?

hard green pear or two each, every year or so. We tenderise them with mallets on an anvil, then gorge.

The peaches were Peregrine and Duke of York, long gone to peach leaf curl, along with the apricot. The mulberry grew quite promisingly for a year or two, but then faded to a halt.

Even the apricot we tried in a sheltered nook by the house thought better of it, especially after being severely and randomly pruned by a certain cow who had no business being there and who subsequently received a round ticking off.

The plums were Victoria, Czar, Oullins Golden Gage, Deniston's Superb, and Early Transparent Gage. Only the Oullins survives, leaning over at about 65° after succumbing to a gale. The sheep clear its lower reaches of leaves every spring. It occasionally produces a couple of coloured pebbles by way of fruit.

Both cherries have gone.

But we still have a walnut surviving. It grows very slowly and steadily, and so far I've not been tempted to beat it: no sign of nuts, despite being planted on a very rich four-hooved source of nutriments some 18 years ago.

The little fig is still with us, too. It's planted in an old spin-dryer drum, and lives close to the wall of the house. We even occasionally get a small, rather woolly, just-edible figlet off it.

One tree we didn't expect to survive the first winter did rather well, and lasted several years. It was a nispera (a sort of loquat, it seems) that we grew from a pip (eyeball-sized stone) in Nottingham. We potted it on into another steel drum, this time from a washing-machine, and put it by the front door, where it caught the full brunt of the weather. Every year it put forth those exotic long ribbed leaves, come frost and storm.

Then one particularly wet winter done for it. Now it serves as a climbing-frame for a honeysuckle that drapes artistically across the front door. Where else should honeysuckle be?

Above the lintel it overlaps with the periwinkle which proliferates in an old Burco, and approaches from the other side. The flowers meet above the little putti, and in season it looks wonderful. Out of season, it drips down your neck while you're trying to find your key.

# What a Beautiful Place, Summer 1993

It's a stunningly beautiful day in late spring as I write this. I'm down in the bottom field, sprawled on the grassy slope, watching the sheep nibbling a living. Ahead of me, and down the slope, is the wire fence we put up three years ago, and beyond that is the steeply wooded cwm – an ancient glacial valley, threaded by a rustling and freezing stream that follows gravity to the Teifi, and then to the North Atlantic.

The trees on the slope are mainly oak and ash with a lot of scrubby growth and razor-wired brambles between. The big dead elms that were here when we arrived eleven years ago have long-since fallen. It's a jungle down there. Some kind of wildlife haven.

Across the valley are the two big farms, their sheep and cattle grazing happily in the sun. Every tree and bush is a different shade of green. It's quite amazing.

There's a fresh breeze. A heron glides past at head height. Pigeons flash down among the treetops, and in the distance one is calling like a dysphonic cuckoo. There is birdsong everywhere. No buzzards out today, though. Normally there are two or three gliding and keening over the cwm, at least until they're chased off by the crows.

A dog is shouting a mile or two away. In the distance is the rumble of a military jet that adds a note of caution. A Tornado, or worse, a Harrier, suddenly passing over at a soul-shuddering 300ft has to be experienced to be believed.

But at the moment, the noise is in the distance. And, surprisingly, there is not a tractor or a chainsaw operating within miles.

It's not often that we take the time to stop and simply

'be' in our surroundings. But it is deeply refreshing when we do. It reminds us of why we came down here. And the world seems even more vivid and vibrant to us today, because yesterday we buried our little dog who'd been a friend for fifteen years. It's a sad reflection that it so often takes a big emotional event to sharpen our antennae. But today, I'm sharper, and see further.

Anne bought two trees and asked which I thought we should plant next to Porky. A whitebeam or a horse-chestnut? Well, there's no contest, is there? Dogs and conkers go together like Kiss Me Quick hats and candyfloss.

I've just become aware of a baby rabbit who's crept out of the cwm. He's sitting dead still; knows I'm watching. There are very few rabbits this year. Myxomatosis has come round again, possibly introduced by a farmer who's sick of losing acres of grass and hundreds of pounds of income to them. So the good news is that there's a lot more grass for the sheep this year. The bad news is that for the second time this week we've happened upon a fox in the farmyard, creeping up on the geese. Short of rabbits, I guess. Or are geese just easier targets?

The goose has been making some sort of half-hearted attempt at nesting, but the eggs keep disappearing. Two of the shells are lying near me on the slope, a hundred yards from the nest, along with the remnants of three chicken eggs; the fox, no doubt. There's also a tiny oak tree, about three inches high. The acorn must have been carried uphill, out of the wood. A jay? I saw one in the garden recently.

If I had Dad's airgun, I could shoot that rabbit. And yet … I hate doing it, especially little baby fluffy bunnies; and there's not a lot of them just now.

Best leave him to take his chances with the fox in the incredible springtime panorama we're privileged to be sharing today.

It must be the loveliest place on earth.

\* \* \*

It's still too easy not to stand and stare enough. I guess we're all guilty of that. We become so obsessed with the small change of the moment that we forget to look at the real treasures of life. Is that column of figures really more interesting or important than the sparrow on the window ledge?

And what can the spites and meannesses of *EastEnders* hold for us that are not stunningly outweighed by even a half-decent sunset? Or the pattern of veins on a leaf?

Just as we are what we eat, so we tend to become what we allow to preoccupy our minds. Anne once had dealings with an accountant who'd had a stroke. He couldn't speak. Well, not quite true: whenever he tried to speak, all that came out was a stream of numbers.

# Radish squash

The advent of the organic growers' co-operative was a big jump forward. Between the dozen or so of us we would be able to market our stuff more effectively and profitably. There was a lot of organising to do. For example, we had to agree on a fair way for us all to chip in to pay for one of us to do the office work. This job fell naturally to Dot, an accountant in her previous life. Old skills can come in handy, eh?

We personally were so taken with the new possibilities, that we splashed out on a third tunnel and erected it in the bottom field. It was immediately christened 'Poly 3' (no shortage of imagination here, mate). It was 60 feet long, twice as big as the others. We now had a total of 1,675sq ft of protected growing space. Excellent.

We bought the third tunnel because we'd been so pleased with the winter lettuces we grew in Poly 2. The Topaz and Cavallona grew slowly but surely over the winter, and were ready for sale to local pubs and restaurants as huge fresh and delicious lettuces weeks before anyone else, and at Easter time, too, 'The Hungry Gap' when we had no other income at all. A nice little earner. Perhaps we should grow a whole Poly 3 full?

We changed our minds about this when another member of the co-op suggested 'radishes' to us.

'Radishes? Tiny little … tiny squibbling, poky little … *radishes*?? Six feet of expensive head room for six little inches of crop? Are you joking?'

But he wasn't. He persuaded us that we could grow an awful lot of top-notch and highly valuable bunches of radish in our posh new tunnel. And he was right.* But we still grew some winter lettuce as well, and whatever else we could squeeze in, like chicory and celtuce.

These were interesting crops, but they never really took off. People tend only to buy what they know or like the look of, and celtuce was an unknown that didn't look quite normal. Thus, it definitely qualified as what greengrocers call 'queer gear'. And queer gear doesn't sell.

The chicory might have turned out OK, but we decided it was going to require too much in the way of time and specialised kit, as it needs to be grown and lifted, then stored and forced. Too much fiddle unless you were going to go in for it big-time.

We were coming to the conclusion that you get the best cash value out of a tunnel by using it to extend the season at both ends, rather than just doing the traditional greenhouse thing of packing it with one crop of summer toms and cues.

Instead, doing it our way, you get a first full crop of early courgettes, say, and a second full crop of late french beans. This means not just that you have a double crop of fresh stuff for a longer period, but more importantly, you have them when nobody else does. The alleged 'Law' of Supply and Demand means that these crops thus pay much better per square foot of tunnel space, than the equivalent foot-week of toms and cues would, particularly as they would be cropping when everyone else is: an invitation to Glut City.

So, doing it our way, you work hardest in the tunnels early and late in the season: once before the mad rush, and

* If you really can't wait, turn immediately to page 158 for more details on radishes. I promise you it's every bit as exciting as it sounds.

again when it's over. During the main cropping months, when you're working flat out, you concentrate on outdoor crops.

And it's nice to be working in a warm tunnel in spring and autumn, but no fun in a full sun.

I'm sure there must be an ideal balance to be found here, whereby you can juggle various earlies and lates throughout the entire season and double your productivity in the process. We were working towards it, but as things turned out, we never quite got there before fate intervened.*

After the Great Pumpkin Disappointment we decided to go for something a little more tasty and a little less unwieldy, that would also be less likely to be gutted, crudely fenestrated, and illumined with a penny candle by a sub-creative sub-manager.

Acorn, Buttercup and Butternut, varieties of American squash, were a marvellous discovery. You can plant a couple at the back of the tunnel bed, right next to the plastic, and just forget about them. They hurtle up and down their run, chucking out leaves like dinner-plates, and frightening-looking tendrils, and at the end of the year, after you've pulled out the various crops and haulms, there's the lovely bonus: a good half dozen radiantly coloured squashes per plant, and none of them too big to handle.

The Acorn *is* roughly acorn-shaped, but graciously grooved and inclined to taper more; about the size of a butcher's hand and a lovely rich dull green.

The Buttercup looks like a very large pork pie, dappled and striated, again in greens.

---

*Ah! That caught your attention, didn't it? Yes, Nemesis is indeed at hand; but not just yet.

Butternut is shaped like a cross between a heavy-duty beige lightbulb and the Cerne Abbas Giant's most obvious assets.

All of them taste wonderful: a bit like roast chestnut, but nicer. We'll never grow ordinary pumpkin again if these beauties are available.

Patty Pan is different. It grows to the size and shape of a large olympic discus, with a neatly deckled pastrycrust edge. The skin is slightly harder than a discus, which makes cooking it something of a challenge. Should one drill holes in it before baking? Or merely wire the oven door shut? I must ask an American one day.

By the end of 1984 we were making plans to put up four more 60 foot tunnels, at the rate of one per year, to really make the most of the burgeoning demand for protected organic salad crops.

# Romantic vs Death, Autumn 1993

I wonder what influence *The Good Life* programmes had on raising public consciousness? Did Tom and Barbara twang a few heartstrings out there in suburbia and make a few more people question the point of their paper-shuffling lives? Every stranger we meet immediately says 'Ah! The Good Life!' as soon as we mention smallholding, so I suppose the show must have had some lasting effects, apart from just being funnier than most others.

But although Tom and Barbara hit our screens a near decade before we moved to Wales, I don't think they really influenced us. We were already actively organic, and were finding plenty of food for thought in the writings of John Seymour. His books were written from a background of personal experience and are gems of exuberance. It seems to be Seymour's credo that every disappointment is an opportunity and every disaster a learning experience. And that with a bit of vision and a lot of good honest labour, you can work miracles, assuming you have a flexible Plan.

I'm not sure what the Goods' Plan was, but I am sure it wouldn't bear close investigation. There is no way you can run a methane digester off two pigsworth of dung, or grow all you need for a year from one small garden. To be fair, though, you just *might*, if seriously aesthetically-challenged, reasonably aim to knit your own suits. Been there; done that.*

Life is different in SitComLand, of course, and rightly so, but if *The Good Life* did raise public consciousness a bit, I wonder if it also worked the other way as well, in prolonging the myth of Cuddly Animals Having a Nice Day Out at the Market?

Tom and Barbara never had to stitch up a prolapsing ewe, or pull maggots out of a struggling ram's backside. In fact, the

---

* Just kidding. No, really.

whole issue of 'What happens to animals?' was more or less avoided, although I do seem to remember that Tom did once threaten to shoot Lenin, his cockerel.

OK, it's a comedy show, but I still think the issue of death and suffering could be faced a little more directly; (*M.A.S.H.* did it).

It seems to me that the fact of the matter is that people don't want to know. They don't want to associate their Sunday lunch with that furry cuddly little lamb. I also suspect that, deep down, most people don't want to have to think at all, particularly about 'difficult' realities, like Death, for example.

This came home to me quite vividly recently. I've been doing some stuff for the local radio station, partly about smallholding. One day, we started recording a piece about geese. Everything was going nicely until I caught a look in the interviewer's eye. He could hear the producer in his headphones. I asked what was wrong. Amidst some polite disclaimers, it became clear that the idea of killing geese would be unacceptable to the station's listeners.

'Are they all vegetarians?' This question brought the predictable polite silence.

I see the producer's point, of course, but are we really all such cowards as to even refuse to listen? It would seem so. I realise that some people are more squeamish than others (and I put myself very near the top of the list) but I do find this refusal to admit the existence of death very puzzling. Is it just fear? Or a slowly developing sense of hypocrisy: that meat-eaters find the promptings of their conscience uncomfortable? Who knows?

But it doesn't do to take all this life and death stuff too seriously. To John Seymour's 'vision' and 'hard work' I would add 'a sense of humour'. Tom and Barbara would understand.

# Learning curve ...

Well, that was fun, wasn't it? Wait till you see what's coming next.*

Death is never far away on a farm or smallholding. When we first arrived we tacked our fields out to various neighbours. This brought in a tiny income, and kept the grass in decent fettle. The following spring Mr Jones from up the hill had his sheep on the bottom field.

One brisk morning, we were out surveying our new rolling acres, and noticed Mr Jones kneeling in a far corner. He was with one of his ewes, who was clearly in a bad way.

'What's up with her?'

'Lambing.'

He indicated with the hypodermic. 'Something wrong inside, see. I tried, but it's all wrong. Now she's been pushing too hard and the lamb's gone down.'

'Gone down?'

'Down into the body.'

'Oh lor.'

'I tried a caesarean, but ...'

The lamb was clearly dead. The ewe was breathing heavily. Soon the anaesthetic would wear off.

Previously, our experience of death was very limited. Even family pets had been 'taken to the vet' by kindly parents. Now it was rasping away in front of us.

'Can you save her?'

'No, bach. No, she's had it, I'm afraid. Yes.'

---

* It's rather nasty, I'm afraid. Please feel free to skip ahead. Page 130, for example, is about runner beans, and I promise you no nasty surprises there.

Then there was a pause. 'I don't like to see an animal suffer,' he said. 'so I do the kinder thing.' And he pulled out a penknife.

There was another pause, as he waited to see what we wanted to do. Stay or leave? I remember my pulse quickening.

I forget now who said what, but Anne and I both knew we had to stay. Mr Jones leaned on the shoulder of the panting sheep and pulled her head back. 'Easy now, bach. Easy.' Then a swift tug on the blade, a brief struggling kick, 'Easy … Easy …' and the head settled back, held gently down by the farmer's arm. The blood flowed briefly, then it was over.

Mr Jones wiped his knife. 'Well, now you've seen the worst of it.'

I was feeling quite dizzy. Not exactly faint, but not far off it. I mumbled something and wandered off to find somewhere to sit quietly. Anne is made of sterner stuff (well, she's a woman, isn't she?) and came and sat by me a little later.

'You OK?'

'Yes. Are you?'

'Yes, I think so.'

'Bit of a shock, though.'

'Yes.'

It was an unexpectedly speedy introduction to the reality of life with animals.

Our next major trial followed a couple of years later, when Star, one of the first sheep we bought, also got into trouble at lambing. We read our books and did our best, but couldn't see how to help. Ken came across and reached inside her. The problem was that the lamb was too big for her. If we did nothing they would both die. Decision?

He did get the lamb out, but broke its jaw in the process. He went behind the barn and dispatched it for us.

Next morning, Ken brought us one of his orphan lambs to put onto the mother, but it didn't thrive, and died as well. So he brought another.

The mum never recovered her strength fully, and then became infected. She died too. Anne stayed tearfully with her to the end. I chickened out. All this mayhem was getting to me, more even than I thought it would.

That left us with the doubly-orphaned lamb to hand rear. He was called 'Simon' and went down in my personal memory as the little cutie who grew up, and got mean, and kept trying to knee-cap me ... until the Day of the Sledgehammer. HA!

Since then, we've obviously had to cope with more deaths. None of them are easy, especially if the beast is a friend: you become very fond of hand-reared lambs, for example, even if they try to smash your kneecaps.

Another difficult one was April's stillborn firstborn. As beautiful as only a Jersey calf can be; born lifeless. Perfect features. Just no life in him. Don't know why.

We buried him in the orchard, in the next allocated space, between the Egremont and the Bramley. Previous plots were occupied by Star and her lamb, and a couple of other baby sheep casualties who had unexpectedly fallen off their perch.

That was the best place for them, we thought. Their nutrients would help feed the trees. We had to dig them pretty far down, though, or cover them with a big stone, or the fox would dig them back up again. Free lunch, you see. Who could resist?

Although we cope pretty well with the occasional death these days, it is still a saddening experience.

One morning, after we'd planted yet another lamb, by the Ribston Pippin this time, we noticed one of the adolescent calves trying to mount another.

Anne pointed her spade: 'The rural idyll of a smallholding eh? All sex and death if you ask me.'

Not *entirely* like *The Good Life*, then!

### <u>Star the Sheep</u>
*As you breathe, at once deep and shallowly*
*chest upping and downing*
*mouth vacant and puffing*
*feet long forgotten*
*only the thread you hang by*
*holding you*
*I watch and stroke your woolly brow a little.*
*Helpless.*
*One last huff and a pause that lasts*
*forever*
*as the gleam mysteriously fades from your green eye*
*and you are gone*
*leaving only memories and wonder.*

# Geese, Winter 1993

Why does anybody keep geese? They're noisy, stupid, messy, stupid and aggressive. And stupid.

As I write, the gander is sounding off. He's not honking. I've never heard a goose honk. The noise he makes is a sort of whistling growl, punctuated by the screech of a pig in torment. His wife agrees with him of course. Oh yes … he can do no wrong. He can pull all the feathers off the back of her head when mating … leave her neck raw … cause her eyes to slitten as scar tissue forms … but he is still Mr Wonderful. Does he protect her? Not that I've seen. He'll make a lot of noise, of course.

I don't mind animals being stupid. I guess that's why they're not all nuclear physicists or estate agents, but I do seem to take objection to the sort of mindless arrogance that a gander displays. The brute outside my window has now begun parading up and down the yard, shrieking. He expects the ducks and chickens to be impressed. I'm glad to say they're not. The ducks just wander past in an extended flotilla, heading for the pond. The chickens are all out to lunch, as usual. But this does not register with Adolf. Up and down he goes, howling and bawling. Perhaps I'll make him a wingband with a little swastika on it. He'd like that.

Isn't it extraordinary how we anthropomorphise animals? How can a reasonably intelligent person even begin to despise a goose for being a goose? And just because ducks have turned-up beaks and look cuddly, we think they're sweet, don't we? But have you seen a duck smash up a slug? Bears are cuddly too, ho ho. Pigs are intelligent (can't say I've noticed). Sheep are stupid (er … ), and cows are, well, bovine. Dogs, of course, understand every word you say. And cats can turn lights on and get stuck upside-down in drainpipes. Thus proving something or other.

Intelligent or otherwise, animals all have their own monomanic gut-focussed agenda, which involves a lot of ME ME ME and very little else: 'sharing is for wimps', writ large. But just occasionally, you get the feeling that some of them do *try* putting a gloss on it.

For example, it's an education putting the poultry away for the night. As I reach the back doorstep, I'm instantly surrounded by an attentive claque of chickens and ducks. I feel as if I should at least deliver a sermon, if not produce a basket of loaves and fishes. They all stand there, or jostle gently round me looking friendly. Of course, all they want is their two dollops of grain, but they *look* friendly. They're *making an effort*. The geese on the other hand just pelt me with common abuse all the way to the grain store and back. As I usher them gently into their shed, with a scoop of oats each, all I get in return is ear-shredding paranoia.

Meanwhile, 'the five thousand' waddle and pad ahead into their own shed. All sweet and cute. But as soon as the grain hits the feeders, all hell breaks loose. Kicking, yelling, biting and elbowing. I've seen nothing like it since the last privatisation shares went on sale.

But a farm wouldn't be a farm without at least a few of these strange little dinosaurs around, which is why I can put up with their appalling table manners and gross incontinence. I can even put up with the gander's ego problems, at a pinch.

What I do find hard about keeping poultry is their mindless vulnerability. We must have lost twenty to foxes this year alone. We feed them, and they feed the fox. We've not eaten a chicken for years. So much for self-sufficiency.

More on foxes another time.

# More than we bartered for ...

We got our first geese in 1983, more or less by accident. They came from another Ken, this time a theatrical scenery-designer of all things (... and cartoonist as it turns out!) who lived in a cottage a couple of miles away. We had previously met socially, and as he looked round our yards he was very pleased to notice a quietly rotting two-wheeled trailer-thing with metal cogs and gubbins at its rear end.

'Oh ... do you want this?'

'I've no idea. Do I? What is it?'

'A muck-spreader. You shovel it in here, and see those bars and chains along the bottom? They shift it to the back and those curved blades on the top (he meant the gubbins) chuck it all out everywhere.'

'It's for a tractor then?'

Did he think I was joking? Was I?

We eventually decided that we didn't really want it as we hadn't got enough muck, or a tractor. Anyway, it was broken. Any fool could see that. I know I could.

'Yes, but I only want it for spares. I've got one a bit like it at home but one of the floor-chains has snapped.'

Well ... wheeling and dealing coming up, then? We've never considered ourselves as General Traders before. I fetch my trilby and camel-haired coat:

'So what's the deal, Ken, me old pal, me old china, Gor Blimey, luvva duck?'

'Well, I just wondered if you wanted it, that's all.'

Oh.

'Do you want any geese?'

And that was how the deal was done. We didn't actually

*want* any geese, but as they were offered, and as any form of possession is better than none to people whose income has dropped by three-quarters, we heard ourselves mumbling 'Er, s'pose so. Er … thanks.'

A day or two later Ken reappeared on his old Ford Dexter tractor, with a big box tied on the linkage, bustlingly full of Adolf and His Adoring.

'There you go. I should give them a few minutes to settle down,' as they emerged amid a welter of shrieks and bright white feathers.

Those minutes extended into a few hours.

So Ken trundled and clanked off with his broken muck-spreader and we had two more Happy Campers. Happy? Those hours extended into days; then months.

Other people's geese never seemed to be as persistently psychotic as those two. A troubled time in the egg, perhaps?

We eventually got another female from another neighbour, and gave them all a corner of the Woodbarn. We fenced it off against foxes, and opened and closed their pop-hole at night. They laid eggs, and began sitting. What seemed like months later, the eggs hatched into brilliant golden little fluff-balls. They were a true delight. Then you looked at the parents and knew it was all going to end in tears.

Actually, they weren't all awful. Most of them never got a chance. They disappeared at night while still tiny. Rats? Polecats? Weasels? The dreaded mink? We did our best at protecting them some more, but never solved the problem properly. My own private theory is that Adolf kept watch till all the females were asleep, then, fangs dripping …

I don't think we'll bother with geese again. They do good

work in trimming grass where you don't want sheep to go, like in the orchard, for example; and the goslings are real jewels. But on the other hand, ours were dreadful boors, and you were never quite sure you wanted to turn your back on them. Even their snow-white plumage, stylishly edged with orange beak, legs and feet, was offset by the icy vacuity of the bright blue eyes. If they spotted an easy target, like a visiting child, for example, the neck would go down, the wings would extend and a bit of brainless bullying would ensue. Occasionally someone got a nip. One irritated visitor fought back and caught Adolf a clout round the head with a well-aimed kick. He apologised to me later, but I reassured him that he'd done the appropriate thing. Serve him right. Please do it again.

I once saw them hassle Anne. She just stopped still, then said 'Boo!' Oddly, it shut them up. So I now know someone who actually has said 'Boo' to a goose.

They were also unbelievably incontinent, and thought nothing of it. The going rate could be measured or timed almost to the degree of one green splat for every five paces. Wellies were *de rigueur*, always. Green ones, for preference. All those squits meant a rapid transit digestive system: fine when grass is free, but expensive when winter comes and feed means expensive grain. As a crop they were not cost-effective. What did we get in exchange for the grain? A couple of eggs if we were lucky, and a tiresome and endless inquisition.

People say geese are good watchdogs. They are supposed to have warned Rome that the barbarians were upon them two thousand years ago, but it never struck me that ours were much good, largely because they were likely to erupt into the full oratorio if a butterfly flew too close, or they saw a rabbit on the yard. I warned them. 'You'll cry "Rabbit!" once too often,' I said, 'and then you'll be sorry.' It had no effect.

The arrival of the postman really freaked them out, as if a totally unknown new terror had descended upon the world, regularly at eight every morning.

Ironically, or karmically perhaps, it was the postman what finally done for Adolf. The idiot was so full of how loud and masculine he was that he failed to notice that the van he was frightening off so successfully was now changing direction and slowly reversing towards him so that the postman could turn round and leave, as he did without precedent *every morning*.

We didn't actually witness the event, but we heard the screeching and carrying on. 'Shouting "Rabbit!" again' was all

I thought of it. But later we realised the normal bellowing had ceased. It had gone quiet, for once. We found Adolf leaning up against the wall of the byre. Gorn.

They say the normal way of killing a goose is to place a broomstick over its neck, just behind its head; stand on the broomstick; then heave on the goose's legs, in order to dislocate its neck. I would love to have owned the film rights that showed somebody trying to pull that stunt on Adolf. They would be worth millions.

* * *

The muck-spreader eventually returned to us. Ken couldn't use it for spares because the parts were incompatible (no surprise there, then), and once we'd got a tractor we realised we might be able to use it ourselves if I could figure out how to mend it. So we swopped it back in return for a hundred visiting cards I knocked up for him on the Adana. It is now sat where it started from, in the bottom yard, still quietly rotting. We couldn't mend it, so never used it. Instead, we bought a more-or-less working model, a year or so on.

# Runner Beans, Spring 1994

Runners have always bean (sorry)* a favourite veg of ours, and we've spent some time developing a quick and efficient growing system.

We tried the traditional methods first of course: the popular Bridge on the River Kwai construction, which involves two rows of parallel poles, all angled over to a central line and tied off in pairs, with a horizontal pole or two tied across the crux of each pair. This system works well enough, but there are two problems with it. Firstly, it takes forever to erect, especially over a 100ft row, and secondly, it suffers badly in a high wind, even to the point of breaking poles. One of our rows fell almost completely to the ground. There was no practicable way of restoring it, so picking the beans was something of an adventure. 'I think I can see one under there … I'm going in. You wait here.' 'Oh, do take care, husband dear!'

To try the second system, the Weaver's Dream, we knocked up the usual goalpost shaped frame, and tied strings to dangle from it for the beans to scramble up. This works quite well also, but it is a fiddle to erect, especially over a long run. And you can't use bamboo for the downposts, as the weight acting on them and on the crossbar is considerable when all the plants are full grown. Thus, to get the strength you need you have to go through the lengthy palaver of digging holes and banging in heavy posts. A considerable labour and expense over 100ft. Finally, it is a terrible chore separating the haulm from the strings at the end of the season.

We refuse to junk the string on ecological grounds, and

---

* Broadly sorry, anyway. Woops.

anyway the cow loves the haulm. If she ate the string as well, it would all get caught up in a ball in at least one of her stomachs and make her unpleasantly and expensively ill; AND ruin the string.

So we designed our own Wigwam City system. It's only a refinement of another traditional system, but it works splendidly and is simplicity itself.

You start by pushing 8ft bamboos into the well-composted soil, 1ft apart along the complete row. You repeat this, to form a parallel row 2ft away. You then push the first four poles of each row together (ie, eight poles) to form a wigwam (or tipi) shape, and tie the crux securely. You repeat the procedure with the next set of eight poles. You now have two wigwams, each covering a ground plan of 3ft row length by 2ft between rows. The finished rows of wigwams are separated from each other by a 1ft space.

That is the basic system. We incorporate an extra pole on each of the 'open' sides of the wigwam. This adds stability and allows for two more plants to grow. We then either plant a sturdy seedling at each pole when the risk of frost has passed, or, more usually, sow two beans near each pole. Varmints tend to eat half the seeds, so the wigwams don't get over-crowded, but even if most of the plants survive, the growing space per unit seems to be quite adequate.

Apart from speed and ease of mantling and dismantling, the wigwam system is very stable, especially with the two extra poles. It also allows air and light (and most gales) to pass through, and the paths between the wigwams allow for easy picking and weeding. We found a Wilkinson 'Swoe' unbeatable for hoeing round poles and inside awkward spaces.

Where an 11ft row of the River Kwai system will give you

22ft of cropping area, an 11ft run of wigwams gives you 30ft (although this area will decrease as the plants reach higher up the poles. But you want to be picking earlier rather than later, don't you?) Or you might want to try running horizontals between the cruxes of adjacent wigwams. This would allow for some overgrowth to continue setting beans.

If you are very fond of runners, and plan on having lots of wigwams, you might want to invest in a sack-tying tool, which will tighten a plastic-covered wire round the crux very quickly and efficiently. It can be a bit of a pantomime trying to hold ten springy bamboos with one hand while noosing them in a length of baler twine with the other.*

And if you like French beans, but are fed up with splashy mud ruining the odd bean that the slugs overlook, why not grow climbing varieties up wigwams? You need less seed, the plants soon grow out of the reach of slugs and mud, and they're a lot easier to pick.

Have a terrific season.

\* \* \*

Re-reading that piece, it strikes me that most gardeners aren't going to be bothered about mud-splashes on French beans: you just wash them, don't you? But for a market gardener, that extra process means hours of wasted time, and a surprisingly complex and extensive array of facilities for washing, rinsing and drying. The beans need to be perfectly dry before shipping, or they rot.

A little three-part brainteaser:

---

* Frankly, I don't believe it can be done. We always worked in pairs on this job. Perhaps this might make a good game for village fêtes? It would make a change from pig-sticking, and ducking the gossip, at least.

1. If it takes a skilled man eight seconds to wash and dry one french bean, using an efficient combination of newspaper, towel and hairdryer, how long does it take to treat a crate of about a thousand beans?
2. At the current minimum wage rate of let's say a fiver an hour, how much extra cost has gone into that crate of beans?
3. Is there a chance in a million that the grower will recoup this expense in the return he gets from the wholesaler?*

Runner beans are particularly susceptible to gluts, we found, which makes them a tricky crop to wholesale. Shops aren't a lot of help either, as everybody with a couple of square feet of garden or a heavy-duty window-box, or even a particularly grubby sock seems to grow their own, and are happy to sell their inevitable surplus to the nearest shop, for washers.

Ironically, we found they're not at all easy to grow as a reliable crop on a large scale. Perhaps we are just a bit too high or exposed here, but we had some serious problems with wind. That's beans for you, I guess.

* Hint: 'No.'

# Of woolies and wallies ...

The last big step forward of 1984 was that we finally got some sheep. You may wonder why we hadn't got some earlier, as this is sheep country par excellence, and we were literally surrounded by the little bleaters, but our Plan was very keen on One Thing at a Time. All the boring infrastructure needed to come first: the allocation of land; the fencing; the tunnels; the irrigation gear; the cultivation equipment.

We'd sorted all that out as best we could, and then brought in the first priority (the house-cow) and given ourselves adequate time to learn enough about her to get by for the moment. The next step was the tractor, again allowing enough time to pay full attention to learning its foibles. What next? A ferris wheel? A helipad, perhaps? Ah, yes ... sheep.

But which breeds to get, and why? Which would do well here? Which might not? Were any of them triploids, or whatever?

Indeed, why sheep at all? Did we need them? Surely the cow and calves would be providing our milk and fertility? We kept reading our books and talking to Ken.

More than anything, we brought sheep in to keep the pasture in trim. A cow wraps its tongue round a swathe of grass and rips it off. Thus it can only thrive on long grass. But a sheep much prefers short grass that it can nibble with its little rows of pearlies. This close-nibbling also takes out a lot of tender young weed-seedlings that might otherwise grow on, at least to dwarf status, and seed in a cow-only pasture.

The usual procedure is to put cows onto fresh grass and follow them on with sheep. That way everybody benefits. The grass is efficiently eaten and fertilised, and if you keep the two species separate, pathogens in the dung of the one are given time to die out while the other species and its peculiar pathogens takes over.

What little I've seen of horses leads me to believe that they must do best on a field that is scoured down to a heavy baize. Heaven knows how they manage, but they seem to.

Our sheepy trio came from a nearby couple who persuaded us that Jacobs were the variety best suited to us smallholders as they are hardy and have good feet (see Foot Note),* lamb quickly and without undue complication, give good meat, and on the aesthetic side, they look terrific in their brown and cream piebald coats. What's more, they clearly put a lot of time and effort into creating a firework display of horns that any self-respecting coven would die for. Some rams have six of them.

On the downside, they are small, give little meat, and their wool is pretty kempy, meaning that it contains more armpit-hair than more modern breeds who have been selectively bred to give huge billows of hairless pure wool.

So Cheeky the Jacob came to live with us for the rest of her long life, along with a couple of chocolatey-coloured Suffolk/Jacob crosses called Star and Rocket. They were all pregnant by a Suffolk ram as part of the deal. We over-wintered them and got to know them a bit.

The most obvious thing of note was that Cheeky was smarter than the others. Was this because she was an old breed, that had grown wise in the art of surviving, while the more 'refined' breeds didn't need to think for themselves any more? She was also more wary. HG Wells' *The Time Machine*

---

* This is nothing to do with any known fetish. It just means that some breeds have tougher feet than others.

came to mind. Did she suspect the motives of us Morlocks, while her dozier sisters just accepted everything we did for them as acts of necessary and expected kindness?

\* \* \*

The policy of One Thing at a Time is one that we'd recommend to any aspiring smallholder. It may sound rather unromantic and boring, but unless you're a dilettante or a millionaire, it really is a sine qua non. At least, it is in our experience.

If you take on too much at once, or put all your nest eggs into one small basket, you're asking for big trouble. We've seen other people attempt these things and fail. We don't know of anyone who's attempted and succeeded. Our own experience with garlic, for all our cautious planning, came close to landing us in deep kaki (a Welsh term). Fortunately for us, garlic was not going to be the one-and-only wonder crop, so we were able to re-group.

One person we knew who got badly caught out had the typical get-rich-quick scheme of the city slicker. He knew that early strawberries fetched a high price. So he rented three acres, spent hundreds of pounds on bought-in strawberry plants, more hundreds on getting them planted, and looked forward to a killing.

Lovely jubbly! … $x$ plants, producing $y$ berries at $z$ per pound equals a small fortune. And twice as much next year, because he'd rent more land. These farmers who rent their land out so cheap are real suckers, aren't they!

It rained, didn't it? Complete washout-job. A couple of grand down the drain.

Actually, I think he'd have been in even more of a mess if it hadn't rained. I don't think he'd geared himself up for the problems of weeding and picking and packing and continuity and transportation and storing and marketing: not to mention pests.

# Polytunnels, Summer 1994

At the same time as we bought our first polytunnel (30ft by 14½ft) Dad bought a little wood-framed greenhouse. The tunnel covered four times the ground area and cost half as much. Dad spent ten times as long erecting the greenhouse (don't forget the concrete foundations and the lengthy treatment of the wood) as we did with the tunnel. And the greenhouse blew flat in the second gale and the polytunnel didn't. I reckon that's a cost/labour differential of something like 40:1, allowing for the fact that we had two people working on the tunnel.

Tunnels are strong, simple and extraordinarily cost effective. Ours have all paid for themselves within a single year. What is more, they are patchable and long-lasting, if treated with respect. This means, in the main, making sure your structure is square, true and tight; and then not letting cows wander in and out, or indeed, anywhere near it. You may need extra protection of some sort, either mechanical or electrical.

As with any building project, you really must start with getting the ground plan right. This is worth spending some time on. Firm foundations, and all that.

First of all, select your site for maximum light coupled with maximum shelter. Light comes from the southerly directions, and, for us, the winds come mainly from the south-west. They may well come from elsewhere, elsewhere.

You will almost certainly have to compromise a little between light and shelter. Personally, I would pick extra shelter over extra light if you have to make a choice. 'Shelter' does not mean a solid wall: a solid structure only causes weird eddies and currents that may be more damaging than the original wind.

The best shelter is provided by a combination of trees and hedges, which will absorb and break up some of the energy of the wind and the damaging sudden buffets. It seems to be the case that a hedge line will provide horizontal shelter for roughly ten times its height.

Common sense suggests that a firm 6ft hedge 20ft away will not block any light worth bothering about, and should break up a big wind enough to take its full force off the flank of the tunnel. It might be worth planting one.

It's a good idea to align the tunnel flank-on to the prevailing wind if you can. The rising curve of the skin will tend to lead the blast over the top. A tunnel end-on to a gale is more likely to get into trouble.

The design of the tunnel means that the ribs hold the skin tight, and the tight skin holds the ribs rigid; but this elegant situation breaks down at the tunnel ends. There you have a sudden vertical, with plastic bunched and pleated and inelegantly stapled and nailed onto a doorframe that is supporting nothing and which may become a source of weakness if not dug in adequately. We cant our doorframes out slightly, by about a foot. This softens the verticality a little, which helps wind pass over more easily, and also gives us another 12sq ft or so of growing space per tunnel.

One final point on site selection: if you can build on slightly rising ground, with the ends facing up and down, you will benefit from the fact that hot air rises. Thus, during the occasional spell of red hot weather, you have a small advantage in automatic ventilation from downhill to uphill. A solar-powered fan would help the air along nicely. Poor ventilation can cause trouble in tunnels, especially if they are long, so grab all the automatic help you can get.

Once you've selected the best possible site and orientation, lay out your ground plan.

This means picking a datum point from which you stretch out lines. These lines should be as straight as possible, to ensure that your ground pegs will be positioned very accurately indeed, both in line, and in opposition. We used an old steel field gate to help us with the right angles. Once your four corner pegs are in, check for true by measuring the diagonals.

Erection is simple: first you bang in the ground pegs (12–15in tubes) and drop the ribs in, attaching them to the central spine as you go. This is relatively straightforward. Then you attach your home-made door frames to either end. This can be amusing for onlookers, but again, is relatively straightforward.

The real fun begins when you start trying to put the polythene skin on. People have been known to come from miles to watch the fun from a deckchair with a crate or two of lager to hand. Advice is freely offered.

The important points are that you must pick a warm day that will allow the plastic to stretch a little, and you must have four people on the job, at least to start with. If this means hauling one or two of the audience out of their deckchairs, so be it. You need a bod at each side and each end. Sod's Law (Polytunnels) states quite clearly that *the unseasonal squall will invariably come from the side with no anchoring bod*.

One side of the sheet is tucked into the trench you have pre-dug, about a foot wide and ditto deep, just outside one line of ground pegs. You then weigh the plastic down with rocks, earth, smashed-up deckchairs, etc. The skin is then stretched reasonably tightly over the spine and the other side is entrenched. Now you only need two people. One to do the work, and the other to do the swearing.

Briefly, there are two stages left. The ends have to be somehow gathered and hauled, tailored and forced to fit round the doorframe, and of course, entrenched. Then the skin must be tensioned.

It's our view that a staple-gun is essential for attaching the plastic to the frame. We've tried nails and battens, but the Uzi staple-gun is king. You grab a swatch of greasy polythene and heave it round the edge of the frame, blasting in staples as you go. An assistant swearer is very useful here, as it is a warm day remember, and you are heaving and straining inside a greenhouse. The sweat is running into your eyes and the audience is smirking and nudging each other. Any dogs present will think you might be interesting to lick. But eventually it is done.

All that remains is to lift sixteen hot steel ribs three inches vertically upwards to tension the skin, against the combined weight of the infill from both trenches, in a temperature of over 100°F. I don't know what the force required is in Newtons or Horse-pounds or whatever, but it feels enormous, particularly as you can't actually position your body so you can get a proper grip. It's like trying to lift a motorcycle by the pedal, while squatting with your face jammed up against a boiling radiator. This is the bit the deckchair brigade likes best. 'Look how his nose is all bent and squashy! Har har ... Oh! He's starting to slobber now, see ... '

But erecting our last tunnel was different. Several thousand years on, I re-invented the lever and, with the aid of two bricks, some shimming, and a crowbar, lifted all sixteen ribs in four minutes.

The audience thought it was cheating, but I can't say I agree.

# Plastic pointers

We were lucky that we had the space and circumstances to align all our tunnels flank-on. None of them was ever damaged by even the severest wind.

I do remember the first few winter storms, though; lying awake at night, listening to the howls and buffeting thuds, wondering if there'd be anything at all left of our lovely new tunnel in the morning. It survived without a scratch.

We designed our doors in two sections of unequal sizes, to allow us to block each six foot square doorway either fully; or more than half; or less than half. This gave us nine different combinations of ventilation; or 12 if we could face leaving the doors off completely. We did try this once or twice, but gave the practice up after seeing what a couple of free range chickens or ducks can do to seedlings. The nine-way combination seemed to give us plenty of options.

We later made up extra frames with the plastic replaced by netting so the air could get through but the pigeons, butterflies, sparrows, pole-vaulting rabbits, etc., could not. Thus the 12-way system returned.

After having all the doors blown out by one gale we thought we might as well go with the blow, and leave the doors off completely if a high wind was forecast, and thus let the wind flush the tunnels through. It never did any damage that we noticed; and it might even have blown out some pre-clinical pathogens.

Just how strong the polytunnel structure is came home to us some years later when we were dismantling Poly 3, and discovered that the ribs on the windward side had a different profile to those on the leeward. Ten years' worth of gales had gradually flattened their curve; but the structure was still as sound as ever.

Polytunnel plastic comes in several varieties, including an 'anti-misting' sort which allows for superior light-transmission. We would recommend not bothering with anything other than the proper stuff, although 'anti-misting' may be a step too far for the amateur.

We once met someone who tried to save a few pounds by cladding his tunnel with builder's polythene. Right from the start, it was cloudy and very inefficient, and because it wasn't ultra-violet proofed, it cracked into a million pieces after the first season, and blew everywhere, like plastic dandruff. He's still fishing bits out of hedges and ditches. So are all his neighbours.

And get the thickest gauge you can afford. The extra few microns don't just add strength to the plastic, but also give more support to the ribs. The thicker stuff lasts longer, obviously. It is guaranteed for something like three or four years, but our experience is that it will last a lot longer if you take care of it.

We always paint the outer edge of the tunnel ribs white, so that the heat is reflected back off the metal and doesn't build up in it and thence begin to degrade the plastic. That's the theory, anyway. Whether it's really worth doing or not, I don't know.

What definitely *is* worth doing is patching any little tears or holes as soon as you see them. You can get special tape, a couple of inches wide and as tough as a lawyer's conscience. It will stop any tear in its tracks.

Such damage is pretty rare in fact. We occasionally find holes bitten at ground level by what must be mice or rats, but there are also bigger holes, definitely bitten or kicked in by rabbits. Once in, these lovable little munchkins set about devouring every blade of greenery, then burrowing and tunnelling, and, by way of an encore, reproducing exponentially. They're very hard to catch, too.

The only other pest is livestock, simply because they are so big. Sheep are usually OK, but if cornered or panicked, a sharp hoof can slash the plastic. A cow, especially with horns, is anathema. We managed to keep Daisy out of the way, but Wally, in a fit of vernal joy, thought what fun it would be to see what happened if he shoved his stubby little horns into that funny white stuff. It ripped, is what. (But the special tape fixed it perfectly.)

A friend had a much worse time of it. His much bigger tunnel somehow got half a dozen frolicking heifers inside it. They never meant any harm, of course. Roger walked calmly in behind them and gently shooed them towards the opposite open doorway. But then someone suddenly appeared in the doorway, right in front of them. This immediately spooked them and the leader turned sharp right and barged straight through the plastic. All the others followed her, giggling.

It was a big expensive sheet to replace.

On the topic of ventilation, I'm faintly surprised that no enterprising genius seems to have come up with a lo-tec device for automatically pushing air through hot tunnels. By definition, you have a heat differential, and where you have a differential, you have a potential for work. That is right, isn't it? You also have a tube; and tubes mean venturis, don't they? Perhaps someone out there has a bright idea?

One thing the tunnels could not help us with was protection from blight.

After the first shock at seeing all our lovely Desiree potatoes devastated by this dreaded rot, we discovered in our second season that this was one battle we were not going to win. Blight just appears to be endemic round here.

We tried different varieties, including 'blight-resistant' ones but they all turned out much the same. Our solution to the problem was a lateral one.*

What's more, the tomatoes we assumed would be safe behind the plastic of the tunnels, caught the pest just as badly. I remember wheeling away three wheelbarrows full of Roma and Marmande tomatoes; all bronzed and mottled with blight; all doomed to be tipped well out of the way.

In fact I tipped them as far from the veg patch as I reasonably could: on the patch of very rough land behind the Woodbarn. Upon reflection, this was quite close to Dad's beehives, and his murderous bees.

Great Heavens! Is it possible that I inadvertently tainted the gentle little souls with blight?

Nah. They'd all have developed crispy dry spots and curled up at the edges, wouldn't they? Then they would all have gone soft and soggy and rotted into putrid little pustules.

Hmm.

Which reminds me: have you ever seen inside a beehive? All those thousands of little cells, humming with life and activity, except for the blocks right at the bottom, where each cell contains a tiny little perch, made from wax and the styles and filaments of crocuses, where the bees spend the night,

---

* For more ghastly, yet possibly helpful, details of Blight on Spuds, see page 221.

gently zizzing and rocking. Oddly enough, this part of the life of a hive has never been filmed.

# Raised Beds, Autumn 1994

For some years now, we've planted our vegetables on raised beds. It's such an obvious method, it is amazing to me that anybody ever grew things in single rows, and absolutely boggling that any gardener continues to do so these days.

The advantages are numerous. For a start, if you plan the width of your bed so you can just comfortably reach the middle of it from the path on either side, you can sow, plant, weed and pick without ever compacting the soil.

There are four other big pluses. Firstly, because you have dug out a few inches from the permanent pathways and scattered the topsoil on the beds where it will do most good, you have two or three inches less stooping to do. A worthwhile consideration if you have a big garden or a bad back.

Secondly, the sunken pathway can be turned into a source of fertility for next year. It is easy to hoe it clear of weeds, particularly if you have a wheel-hoe. You can leave the weeds to rot in the pathway, and chuck any de-soiled weeds from the beds themselves into the paths as well. You now have the makings of a long thin compost heap, ready to be returned to the bed the following season.

This year we're experimenting with newspapers tucked into the 'trench' of the pathway to act as a mulch against re-growing weeds. It doesn't look very pretty at the moment, but it will soon weather down. We'll also try cardboard, straw and sawdust.

The third advantage of the bed system is that you can plant at a much greater density than you would ever have thought possible, in staggered rows, with equidistant triangulated spacing.

For example, we plant onion sets 6in apart each way, forming eight rows up a bed 45in wide. A 40ft long bed holds about 650 sets. Last year (a good onion year) the crop averaged 8oz each. That's nearly 3cwt (about 150kg) of onions; that's about 2lb of onion per square foot. And the only reason we plant them so far apart is so we can hoe (using a 'Swoe') between them. They could be planted 4in apart and would produce an even heavier crop, but of smaller onions.

Obviously, most gardeners are not as obsessive as a commercial gardener is about optimum use of land, but it seems to me that if you can grow 200 per cent of your normal crop for 150 per cent of the work, then it's daft not to do so.

If a small garden has, say, a 6ft by 4ft patch allocated for onions, the gardener can choose to either plant the old-fashioned way, in rows, or the rational way, in a bed. Planting 6in apart, in rows 1ft apart, he will expect to harvest about 48 onions. Planting 6in both ways, he'll get 96. There's no competition, is there?

I know for a fact that at this point some dear soul will say 'But I don't want 96 onions. It's too many.' To you, sir or madam, I say 'Enjoy the sunshine, for the day is short.'

Similarly, we plant eight rows of carrots up a 45in bed. They grow to full size perfectly happily, spaced along the rows as normal, which means 'virtually touching'. They grow so densely that weeds soon don't get a look in. The crop is huge, and of superb quality.

Beds are ideal for planting courgettes at virtually double the recommended density: ie two staggered 18in rows per 45in bed. What is more, it is easy to lay a black plastic mulch across the whole bed, tucked in at the edges, ready for transplanting the seedlings out into pre-warmed soil. The plants thrive, and crop a week or more earlier.

And what are the drawbacks of the bed system? Very few that I can think of. Some crops are not best suited to it, perhaps. Do sprouts really prefer compacted soil, for example? And what about earthing up potatoes? I've never tried it, but I suspect a 45in bed would cope well with a double row of maincrop, or even a triple row of earlies. You wouldn't need to earth them up if the black plastic kept the light off. But ... you might have trouble with voles or mice. We had one particularly annoying season when you could see the cute little voles' shock waves as they dashed along their runs under the safety of our expensive plastic, spreading the good news about all those lovely tasty tubers going for free. Grr. It only happened once though.

Beds take more initial preparation, of course. But once instated they're much easier to maintain, because the fourth great advantage of them is that they don't need re-digging year after year. Normally, a quick lick with a hoe or a brisk raking will bring them back to seedbed status.

One final minor advantage is that beds instantly delineate your boundaries for your three, four, or 25 course rotations.

Beds are here to stay. Have you tried them yet?

* * *

I wonder if anybody actually did give raised beds a try after reading that article? I wouldn't be too sure, as people are slow to change old habits, and I suspect that gardeners and farmers are particularly so.

'I grows my radishes six feet apart, just like my old grandad did. Never did him no harm. Lived till he died, he did. Trod on a rake one day, up she comes, and knocked his head clean off. Six feet apart, that's the way to grow radishes.'

This grossly paranoid and complacent conservatism obviously does not apply to you personally, dear reader, but I bet it applies to somebody you know.

'Turnips? Never touch 'em. Them wossname Romans brought 'em in, din't they? Foreign muck. Potatoes? Runner beans? No, thank you. No way. I'm stickin to pigroot and hawthorn berries; proper British food.'

I was once astonished to watch a television programme about a cattle farmer who was facing ruin from BSE, but who seemed quite unable to contemplate the possibility of farming some other way. 'Sheep?' He looked blank. 'Mixed? Field crops? Herbs? Nature reserve? Organic?' The poor man just looked panicked at the very thought of non-cow farming. It had been in his family for generations, you see. 'Change?' He almost did not understand what the word meant.

In the end he did change, of course, because he had to. And I've no doubt he'll do well, once the idea sinks in. But farmers and gardeners do tend to be a conservative lot. I guess that's why governments feel they have to dish out so many subsidies, as being the only way to bring about any sort of change at all. We're certainly not seeing a lot of dynamism among farmers over the switch to organic growing with its guaranteed and expanding market. Have they become subsidy-dependent, in the same way that their souped-up cows have become factory-feed dependent?

The most famous illustration of bucolic conservatism that I know concerns the period when potatoes first came to France. The king (Leroy?) realised that they would make the ideal staple crop for his hungry and incipiently revolting subjects. But he knew their conservatism would never let them grow them, even if they were given samples for free. His solution was ingenious. He grew royal plots of spuds and let it be known that these marvellous and delicious fruits were much too good for the hoi-polloi and that they would be kept under armed guard, day and night. Secret orders ensured that the armed guard nipped off for a fag, or possibly a baked potato, now and then, and didn't hurry back.

Within a few days the potato field was cleared, and the purloined tubers rapidly spread to become the staple diet of the French peasant. Clever old Leroy, say I. I wonder if he casually let slip that they were aphrodisiac as well?

# That's the way to do it ...

. field scale, the only sensible way to make up beds is with a tractor. As I got used to handling the Fergie I realised what a marvellous maid of all work it can be. Making a 200 foot raised bed by hand would take all day; probably longer. The Fergie could do it in minutes. That's not strictly true, but it certainly cut the work by three-quarters or more.

The tool we used was a battered old Fergie ridger we bought for a fiver at a sale. This is a device that rides on the hydraulics and looks a bit like a rack of three double-bladed ploughs.

It works on the same principle as a plough. The snout digs in, and as it moves forward, the soil is forced upwards and to the side by the curved blade. Each pass up the field produces two complete furrows and ridges, and begins a third one which you use as a guide line for your next pass. Brilliant.

If you're clever, you can plant spuds in the furrows, then drive back up the rows, half a furrow out of phase. You can now lower the three units into the tops of the ridges, causing the blades to split them back to where they came from, and, thus, burying the spuds. Really brilliant.

I tried it twice and gave up. Some ridges split and some didn't; I drifted off course and ploughed the spuds back out; I over-corrected and ploughed them back in again, but in a different place. It was a real mess. We had spuds come up all over the place next spring.

I discovered years later that I was lacking an essential bit of tack called a stabiliser bar. This simple steel strip locks the ridger to the tractor good and rigid ... otherwise the ridger drifts and wanders about on its linkage, making accurate work impossible. I took a small crumb of comfort from that.

But we didn't need a stabiliser for making the beds. We decided how wide we wanted them (about 45 inches, which neatly coincided with the track of the tractor) and set up the outer units of the ridger accordingly. The middle unit we removed.

Thus the outer units churned soil inwards to the bed between the wheels, and outwards into the beds-to-be, to left and right. All we needed to do was shovel the furrow out a bit, to make a decent walkable pathway, then rake the beds into a fairly level surface. A nice little 45 inch hydraulic-mounted cultivator would have come in handy here, to make a one-pass seed bed, but they are very expensive and our scale of work didn't merit one. The Howard was fine for us.

\* \* \*

The experiments with strip composting between the beds worked rather well, although we had to be careful not to overdo the paper (as with any single component in a compost), and wind was an ever-present nuisance. Until it was good and wet, sawdust and straw blew about a bit; so did newspaper. In fact dry straw was pretty hard to keep in the pathways at all. Cardboard was a real fag to tear up, and we never quite worked out a time-effective way of dealing with this one. Soak it? Too heavy to shift when wringing wet. Strim it? Noisy, messy, and probably takes more time than tearing. Chop it? Same problem. But there must be a way. Perhaps one of those old Victorian cast-iron forage choppers?

I reckon this strip composting is a system worth pursuing, especially if you have access to a load of spoilt, and preferably pretty wet, straw or hay.

\* \* \*

People sometimes question the organicity of using plastic, especially as a field mulch. We pondered the pros and cons and came to the conclusion that if used responsibly, plastic mulching counts as 'appropriate technology'. 'Responsibly' here means 're-used as often as possible and then properly disposed of'.

In fact the plastic we used is incredibly tough stuff. It is actually very hard to poke a finger through. It just stretches into a longer and longer witch's hat.

It can be used for at least a decade of hard mulching, we reckon. It's a fag to lift at the end of the season, that's all.

How does one tidily fold a 200 foot strip of muddy plastic in a gentle breeze? One doesn't, frankly. One begins to, but then one loses one's slimy grip at a critical moment and ends up rushing round the field like an olympic ribbon dancer in claggy wellies and a sensible coat, bawling unreasonable threats and instructions at the cracking flapping sheet.

We did try inventing a simple device that would let us lay (and, we hoped, 'un-lay') this helpful plastic from a roller hung from the back, or possibly the front, of the tractor. It sounds like a simple problem, doesn't it? You drive slowly along, and the plastic just ... unrolls and lays itself, neatly tucking itself in at the edges. That's all.

If you have an hour to spare in the pub tonight, or are desperate for some new way of breaking the ice at a dinner party, now that juggling with flaming flambeaux is so passé, just try turning the conversation towards inventing such a machine, bearing in mind that it must be made from easily available materials, like bamboo and lath, by people with only moderate fabricating skills: ie, those of us who are capable of turning a self-assembly vanity unit and wardrobe into a bedroom shed. All submissions welcome.

More on my efforts at DIY some other time.

# Brambles, Spring 1995

'Oh, what a lovely patch of blackberries you've got!' is one of the nastiest things you can say to a farmer. 'Blackberries' they may be to you, representing free fruit and butterflies, but to a farmer they are 'brambles' and a cruel reminder of neglected land.

Through a rose-tinted memory I can see our first season here, twelve years ago, running in slow motion across the field, the women in Oklahoma gingham, the men in union shirts with the sleeves rolled up and stout brown boots with tags. We laughed gaily, tossing our shining locks and swinging our honest trugs in the sunlight. Eleven species of butterfly we saw, and picked 30lb of fruit.

Then I remember the parallel reality of scarred arms and septic fingers, and the hard fact dawning on us that a sixth of an acre of brambles, out of a total of five acres, is a lot of lost pasture. What's more, the thicket was a perfect redoubt for the battalions of rabbits that were devastating the rest of the grazing. Our neighbour, Ken, clinched it for us. Five rabbits, he said, eat as much as a sheep, and five sheep eat as much as a cow. It was clear that there were at least twenty rabbits out there. Four-fifths of a cow. The thicket would have to go.

Our technique was primitive but effective. John (our WWOOFer-in-residence) heaved the tangle back to expose the root, using the biggest rake we could find, while I hacked away with a 'Turk' scythe. This tool, with its short and lethal blade was perfect for cutting bramble.

Unfortunately, it was not perfect for using as a pusher or lifter or tamper, and after many hours of gross abuse the handle bent, then wobbled, and broke in half. But Alun the

Mechanic welded it back together in seconds, using an old file as a reinforcing fillet.

We bashed and flattened the first patch into as compact a mass as we could, then stuffed it with newspaper and torched it. It flamed and flared like the Reichstag and reduced to ashes in minutes.

Obviously the stumps would re-shoot in the spring, so we tried hacking them out. Squeezing blood from a stone or an apology from a cabinet minister is easier than hacking out brambles, believe me. After a few minutes we abandoned brute force, and fell back on intelligence: what do sheep like best? Tender young shoots. What will bramble stumps produce in the early spring? Tender young shoots. End of problem.

We moved swiftly on, sacking and burning all before us. The last stand was the biggest, some ten yards square. It was on the steepest part of the slope, and gave onto the wooded cwm a few yards below.

We made a huge ball. It lit at a touch and flared like a Bessemer. Then the wind got up and the whole immense fire-ball began to roll downhill, a wheel of fire 20 feet high, heading for the woodland.

For a full ten seconds I had visions of Dyfed burning to the ground, and it would all be my fault …

## Local Man Shackled in Hague: Admits Playing with Matches, Genocide, Etc.

It was OK, of course. The woodland was too damp to burn. This is Wales, after all. There was just a lot of impressive crackling for a few minutes, then it all died back to wherever it is that fire comes from. There were even some unburned nub-ends left, which we ignored.

So … success. We reclaimed a lot of land; the rabbits were forced back ten yards, and, yes, the sheep ate off the tender young shoots, as planned.

But the following year the ex-bramble patch was a mass of bracken …

\* \* \*

Those brambles left for good, but the bracken is still a nuisance. If we remember in time, the best solution seems to be to go on a dawn patrol in the springtime armed with a suitable tool, and cut down every juicy little crozier as it appears above ground. A Swoe, or better still, a golf club, is an ideal instrument. Just slip on the Pringle and pretend you're an eccentric.

A scythe is over the top.

A hammer is just plain vindictive and indicates a serious lack of grip.

# Just hang on in there ...

By the end of 1984 we were settling into some sort of pattern. We'd got our infrastructure worked out and had imported the most important functionaries for the holding, both biological and mechanical. We were growing pretty much all of our own veg, and one or two extras beside, including an erratic supply of eggs from the Holland Blues and a couple of mongrels ('mongrel chickens', I hasten to point out) we'd acquired from new friends and neighbours.

The biggest 'extra' was dairy produce.

When we first got Daisy she had recently calved. Her calf had just been weaned off her, which meant that she was still producing an almost full head of milk. It never ceases to amaze me that this moderately sized beast could produce so much of the stuff. The first foaming bucketful took me back to the moment when philosophy first crossed my path as a nine-year-old at a Liverpool pantomime. I was deeply struck by one of the lyrics we followed the bouncing ball to:

'Oh ... why does a red cow give white milk, when she only eats green grass?'*

Daisy didn't look capable of producing much of anything, let alone milk. For a start, she had the bony haunches that all Jerseys have, and legs inclined to super-model skinniness, and let's face it, something of a pot belly; well, on one side anyway. I learned later that all cows are built like this. The side with the first stomach, the rumen, seems to be almost infinitely expandable.

---

* This point has never been adequately answered by the world's philosophers; indeed, it is scarcely ever addressed at all. Why is this?

Nevertheless, skinny-looking Daisy would be waiting peacefully at the gate at the end of the cat-walk for us twice a day every day, her udder bulging like an oil-man's wallet. Sometimes she was so full she'd be dripping, or even squirting as she walked. I suggested a new washer, but she looked askance.

Ken showed us how to milk; and didn't it look easy?

The secret is to first get yourself and your cow comfortable. This means tethering her (for *your* comfort) and giving her a dose of whatever she fancies to nibble on while being relieved of her burden (for *hers*).

You may like to offer her a nice armful of aromatic and herbal hay on the one hand or a scoop of 'nuts' (bran pellets enriched with oil, molasses, diseased sheep-brain, etc) on the other. Which one do you think she will pick, if given a free choice in the matter?

No surprises here, then.

Do you know one single child who prefers a nourishing wholemeal sandwich to a saucerful of Smarties? Or a husband who will, of his own accord, select a tumbler of Perrier Lite over a double-handled tankard of Barleywine Five Star Triple-Overproof Heavy?*

While the cow is guzzling and snortling away at the front end, you produce your marge tub with a damp sterilised rag in it, and wipe her udders over. Believe me, this is not in the least bit erotic. She may enjoy the titillation, if that's the word, but you just get on with it. The aim is to remove anything nasty that might otherwise drop into the milk. And 50 per

---

*Or, come to that, a woman who will go straight to the 'Size 16 and Over' rack? (I know that's not strictly equivalent, but I thought I'd better balance things up a bit, somehow.)

cent of the time there is something really nasty to remove. Why is it, that given a whole three acres to frisk and frolic in, that when a cow decides the time has come to just sit and ruminate, she picks one of the few minute spots she has recently voided her enormous bowel onto? Flop. Down she goes, nice warm udder straight into it. And doesn't it dry on well, to that nice warm udder? Like unto a blanket.

So ... you chip and hack and peel and wipe off as much as you think necessary, and put your filthy rag back in the tub.

Then you wipe a little vaseline over each teat. This has nothing to do with suntan lotion or baby oil; it is to help keep the skin supple so it won't crack and bleed and suppurate. Oh, yuk. OK; we're ready.

Now: drop your three-legged stool in *precisely* the right place, and swing the bucket round to end up between your knees, just this side of the udder. We bought a special big shiny stainless steel job, but an old paint tin would probably do, as long as you're not too fussy.

Three possibilities await you: either the whole process will go without a hitch, and ten minutes later you will be carrying a heavy bucket of rich milk into the kitchen to be chilled for processing; or more likely, one of the legs of your stool will begin to slowly sink into the uneven straw-dung litter on the floor. This will unbalance you one way or another. You may end up with your head pressed to her bulging convex flank, halfway to her head, straining to squeeze and pull at arms' length; or you may slide slowly backwards until you are actually hanging grimly on to her teats to stop you falling off altogether; or, if the God of Hand-Milking and Music Hall so wills it, you will inexorably tilt towards the exhaust end. This is usually the cow's cue to fire up the cabaret: with a deft shuffle of vertebrae and hips, she will kick over the bucket, knock you off the stool with her tail, and pump out several pounds of smelly glop, as near as possible to your sprawling form. The co-ordination is remarkable. So is the capacity of that cloaca maxima.

But, as they say about Life, 'if you can't take a joke, you shouldn't have joined'. You pick up what needs to be picked up, wash or at least wipe what ought to be sterilised, relocate your stool and try again.

This time you start to slide in another direction. Never mind. Just keep going.

A gentle thump upward on the udder, to mimic the calf butting for service, followed by a gentle pull and squeeze; pull and squeeze … DOWNWARDS, unless you want to be deliberately kicked clear off the stool again.

If she's in the mood, the milk starts to jet and foam into the bucket. You empty the two quarters on this side, then lean through and empty the two udders; 'others'.

As you stretch your arms under and through, your cheek is now pressed hard into her flank, and you hope she doesn't pull any of her small repertoire of amusing tricks. She might. Her hip can kick sideways as easily as backwards and forwards. That bucket is at risk until you've got it outside.

Some days, she's just not in the mood. Probably a headache, if I know anything. You go through the whole ritual, but she won't release a drop, even though she clearly needs to. Who knows what bovine fantasy she's acting out?

'ME COW. ME BOSS. YOU FARMER. YOU RUBBISH' (RAISE TAIL: DELIVER PROOF).

The only solution is to try again later. She has to give way eventually.

## Cow Explodes Shock! Milk Everywhere, Says Bewildered Farmer

Then she gets her wholemeal supper, and is turfed out for the night, ready to go through the whole performance again in 12 hours' time.

And again.

And again.

Not much of a life, is it? But cows seem to like it.

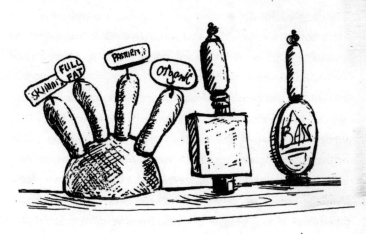

# Radishes, Summer 1995

The humble radish is a market gardener's nightmare. This may come as a surprise, because everyone I know started gardening on a Special Patch Just For You next to the roses with a crumpled packet of Cherry Belle and a Postman Pat watering can. What crop could be simpler? Radishes grow quickly and reliably and can be sucked and choked on straight from the mud.

But it is this very speed of growing which is the problem for a grower. He doesn't want a handful here and a posy there, but a constant 100lb a week, minimum. This means he has to sow regularly and make sure the growth is unchecked and regular too. But the weather is not regular. A couple of hot dry days will check the crop, and a sunny week is a drought. So cropping outdoors is very risky indeed unless you have a good irrigation system.

Our irrigation is less than perfect, but we tried anyway. We sowed the seed (good old Cherry Belle) with a pushalong seeder in 8in bands, calculated to produce the required 100lb of fruit. A week later we sowed another band, and so on. But it was very hit and miss. Apart from the weather/water problem, the seed was not F1 hybrid, and was therefore unpredictable. And as the depth of sowing was very irregular in our stony soil, we would find skinny little jobs next to monsters the size and hardness of champion conkers.

The other problem was flea beetle which appeared from nowhere by the thousand and devoured seedlings back to the stem in hours. Lawrence Hills' Flea Beetle Trolley is an ingenious tool but it can only catch the beetles that jump upwards. It can't catch the critters that eat the soft and juicy underside of the leaves.

I made two wheel-less variants of the trolley that passed over and under the leaves at the same time. They worked pretty well, and saved a couple of crops from certain devastation, but they required a very steady hand and were difficult to clean and service.

All in all, we soon discovered that radish cropping had to be done in a polytunnel, where we could have much more control. Elsoms, the big commercial seed supplier, kindly sent us a seven-page manual of instructions (for radish??) and away we went.

And, yes, it was much better. We used special indoor varieties and kept them well watered. They flourished, and we picked them by the fistful.

Success! The problem came, as ever, with the marketing. In the age of pre-packs, shops don't want bunches which wilt before lunchtime. They want plastic bags, gaily labelled, long-lasting and selling for four times as much, or more if possible.

More profit for the grower? I hear you cry. Well, no, actually. Pre-packing means washing, topping, tailing, weighing, bagging and labelling. And it takes hours, believe me, to process 100lb. All the extra 'profit' is soaked up in the hours of mindless labour.

But there was (and still is, I would imagine) an insatiable demand for pre-packed organic radish. I calculated that two people working flat out with the use of a couple of dozen big polytunnels, appropriate machinery and (cheap, exploited) labour, could become millionaires inside fifteen years or so, selling to supermarkets.

But apart from the fact that monoculture is inorganic and does not appeal to us, we knew from experience that the supermarkets would ditch us overnight if they could buy radish 0.0001p cheaper from Korea or Peru. Not a good risk.

Our couple of little tunnels were nothing like big enough to produce even a fraction of what a supermarket would want to be bothered with, even as a market-testing experiment, unless we could combine our produce with that of many other growers, so we eventually packed radish in as a serious commercial possibility.

Now we grow a little patch of Cherry Belle next to the roses. But we do use a proper grown-up watering can.

\* \* \*

Those field radish were a real trial. It felt as though everything was stacked against them growing 'properly', by which a grower means of course 'of uniform size and quality, and totally unblighted by weather, ill, or pest'. A logical and reasonable lot, growers are.

The biggest problem was the flea beetle. They were to radish in Wales what locusts are to everything in Africa.

I made a rough version of Lawrence Hills' trolley, which is a little pushalong device that rattles the leaves of the radish, thus encouraging the beetles to leap up off the leaves and onto a wooden plate which is liberally smeared with oil or axle grease.

But it quickly became clear that wheels were no use on our stony soil, and anyway, most of the beetles were jumping down to earth and safety and not upwards to danger and a Castrolubey sticky end. What I needed was something that would catch both upward- and downward-jumpers.

I fiddled about with various designs and eventually settled on one that could be easily made from scraps of wood, and looked rather like a section of square downspout, about four inches in section and a foot long. One side of the length had a one inch wide slot cut along it. On the opposite side to the slot I fixed an old broom-handle which would allow me to hold the device in front of me at the same angle as a metal-detector. The inside surfaces were smeared with grease, as were my hands, trousers, face and glasses, by the time I'd finished.*

To use the device, you just walk along the row of radish, ensuring that the plant stems pass through the narrow slot, while the leaves pass through the four inch square body of the 'tube'. Can you foresee any operating problems with this design? Surely not?

Oddly enough, it worked rather well; but not all of the time. It depended, of course, upon the uniformity of growth of each radish; and 'uniformity of growth' is the bugbear of any attempt to mechanise horticulture. Anything over four inches across was going to get smeared with grease. Some did get smeared, and were indeed badly damaged; but not as badly damaged as they would have been by the beetles.

The real problem was keeping the hand steady. Once the device was engaged, by sliding it carefully round the stems of the first plants at the bottom of a row, you could not disengage until you'd covered the whole 200ft. I never managed to get to the end without brushing a dozen or so plants. I began to feel more like a minefield clearer than a gardener as my muscles began to cramp up with the effort of maintaining absolute steadiness. Would one error set off a chain reaction? Steady … steady … just 80 more feet to go … steady … Boom! Oh no! Boom boom boom …

---

*By what disseminative process does grease always end up on the *inside* of the lenses of your spectacles? And the handle of your toothbrush? And, in similar vein, how does honey invariably transfuse through the glass of the jar, rendering it permanently sticky? I am aware of Murphy's Law (quoted later when considering bees) but theory alone is never enough. Has this micro-osmotic process ever been observed by an accredited physicist? We should be told.

# Local Man Found Dead in Field:
## Studded with Radish Tops Say Paramedics

But yes, it did catch the beetles that were eating the undersides of the leaves. I added a little springy 'leaf-nudger' at the front, which made sure the plants got a good shaking, and caught thousands of the blighters.

It was so successful I made a six inch model too, so I could keep up with the hygiene as the radish grew on.

If we'd kept on with field radish I'd have carried on with the R & D of the Smallholder's Patent Radish Hoover, and made it less time-consuming to clean and re-charge. Something to do with tinplate and hinges should do it.

I've still got one of them which I sometimes take to talks at local gardening clubs as a 'What on Earth is This?' exhibit. Nobody's guessed correctly yet.

# Your first million ...

If anybody out there is interested, it's still a virtual cast-iron certainty that you can make a million growing and selling pre-pack radish to the supermarkets. You would need, say, a dozen or so 60 foot tunnels. You would also need to be extremely well-organised, very hard working and dedicated, and very very fond of radish. But it can be done, I'm sure.

You would also need to pick a couple of other crops to alternate with the radish. One good contender would be those Ishikura salad onions. Perhaps a Little Gem style lettuce, or a radicchio for a third crop?

Once you've decided on which three or four crops your wholesaler will be sure (ha!) to buy off you to pass on to the supermarket, all you need to do is work out your cropping rate per sq ft of tunnel, and sequence-sow accordingly, so as to be sure you'll be picking 500lb a week or whatever, every week, without fail. You'll have got all your fertilising procedures worked out, and the irrigation will be foolproof. Your varieties will have been settled in advance via discussions with the supermarkets and wholesalers, and those very nice and helpful people at Elsoms, or one of the other handful of commercial seed suppliers.

It will be greatly to your cash advantage to do all the processing yourself, which means having the labour and facilities close to hand for all the topping and tailing and washing and weighing and packing and labelling that has to be done. It is possible that at least some of this must be done at the wholesaler's packhouse, for a number of reasons, including hygiene control. But the more you do yourself, the greater Value Added benefit you get.

hen you'll devise a way to creatively dispose of all the ra... h tops, via either a very patient pig or a very hungry cow (ours refused point blank to eat more than half a bucket at a time),* and you're quids in.

Meanwhile, obviously, you are repeating this process for the onions and the lettuce. They will crop at different times and at a different rate from the radish, and will need quite different facilities for washing and packing, etc. You will discover how incredibly tough onion roots are to cut through cleanly, and will be driven to the brink of drink and beyond at the prospect of having to trim another two thousand before bedtime, and will think seriously of investing in some sort of bench-mounted guillotine or, preferably, a socking great laser.

Then you will need a large reliable van.

The organic side of the operation might be tricky. The amount of dung compost you'll need to keep a dozen tunnels fertile will be considerable. So you might want to be over-wintering half a dozen beef bullocks, say. This requires a suit-able barn, with adequate plumbing, along with fodder storage space for five or six hundred bales of hay and several hundred of straw. Then you'll need the gear to shift, process and spread the muck and then the compost. It's all getting rather compli-cated, isn't it? Never mind. The up-side is that you'll be able to sell the grown-on cattle for a decent profit, with any luck at all, unless they go mad from spontaneously eating a diseased sheep, which is admittedly pretty unlikely, or catch something small, nasty and lethal.

Alternatively, you might buy in your fertility, either off

---

* Strictly speaking 'half a bucketFUL' at a time.

someone else's farm, in which case you are robbing them of their own replacement fertility, or in bag form, which is of dubious organicity, for several reasons. Whose land are you robbing, again? Is the stuff from a guaranteed organic source? How much pollution have you caused in importing it?

It's much more 'organically honest' to grow your own muck, complications and all. But it all takes time and work.

And on the 'robbing fertility' front ... did you grow your own hay and straw?

* * *

So, we sort of had a chance at making big money. We were tempted, even. But as we faced the prospect of putting all our tight resources into one very unreliable basket, we took a few moments out, with a glass or two of Elderberry Knee-Loosening Fluid to stop and ask ourselves 'Is this what we came down here for? To become slaves? To what end? To make a million? And what would we spend it on? Buying somewhere peaceful? Perhaps a little smallholding somewhere? Where we could grow our own stuff and er ...?'

But ... maybe someone out there fancies the challenge. You would need to be dedicated, but it could be done. Any takers? And on the theme of pre-packs and inspired marketing, Dwlalu Farm is pleased to unveil its Autumn Range:

✳ *Gift-wrapped Sturon 'Heart o' Gold' onions individually sheathed in shiny laser-cut silver twists, with 5mm ribbon in choice of eight different tartans!!!*
    *ONLY £4.99!!!*

* *Individual 'Diamond Gem' lettuce hearts, lovingly washed and pressed, with just a hint of balsamic lime 'n' lemon salsa drizzled into the breathable kevlar 'safety-first' bag!!!*
  *ONLY £4.99!!!*

* *Graded sets of six 'Cherry Belles, the Queen of Radishes', star-trimmed, strung onto quills of lemongrass, rainbow-tinted and rainbow-packed!!!*
  *ONLY £4.99!!!*

* *And don't forget to try our new Fun Edible Jewellery range!!!*
  * *The 'Goodness Greasious Me' Onion Bargee Emergency Ear-rings!!!*
  * *The 'Miss Piggy' Pork Gristle Spicy Amber Necklace!!!*
  * *And the ever-popular Thai-style 'Tubba Shugri Phat' wholefood quick-snack hi-energy delight!!! Now with rip'n'drop lid and platinum-effect spoon!!!!!*

  *ONLY.........£9.99!!!!!!!!!!!!!!!*

# Hay, Autumn 1995

As you cruised through the countryside this summer, admiring all those contented cows, I hope you spared a thought for the frenzied panic that seizes every farm and smallholding in the land round about Wimbledon Fortnight, and which enables all those happy cows to survive the coming winter. This panic is called 'Haymaking'.

Good hay is wonderful stuff, with a fresh heady aroma and a dull green colour which shows that the goodness in the grass leaf has been retained and has not been bleached out by lying too long in the sun. It should also contain a fair sprinkling of meadow flowers. All grass and no herbs makes hay a dull chew. Ask any house-cow.

## Talking Cow Witchcraft Horror: Entire Herd To Be Burned At Steak

As a safety-conscious rule of thumb, it takes roughly a hundred bales of hay to overwinter a cow; more if the hay is stalky, due to being cut after the grasses have seeded; and a lot more if it's been baled too green or too damp, when its inner depths will be filled with a choking white mould that causes 'farmer's lung'. Cows don't like it either, and demonstrate the fact by nosing and tossing it around until they find an edible bit. More dust. And more waste: you need to buy in significantly more bales of duff hay to do the job of a hundred bales of good stuff.

So every farmer and smallholder wants to produce the best possible hay, especially as it costs fifty per cent more to shift 150 bales of bad hay than to shift 100 bales of good. And anybody who's loaded, transported, unloaded and stacked 100 bales of hay on a hot sticky day would much rather not shift 150.

It's the weather that makes everyone anxious. Good hay needs a fair spring for the grass to grow plenty of leaf, then a break of five or so hot dry days at the end of June. Five clear days doesn't sound a lot, but it can be a rarity in Wales. Around the middle of June farmers get edgy, devouring every weather forecast with a focused intensity normally reserved for *Baywatch*.

Most years we get a three-day break, followed by a day or two of showers, meaning that the hay, which has nearly dried, gets wet again. It doesn't matter too much if you can dry it again quickly, but some years you can't. This leads to the sad business of baling rubbishy hay. It's wet and heavy, takes longer to shift, and is worth very little to man or beast at the end of it all. Many farmers in Wales now make silage* rather than hay precisely because you only need one or two good days for it.

But this has been an admirable year for hay, and everyone has been working into the night, cutting, turning and baling. The tractor engines can be heard for miles around, and their lights stripe the gloaming fields.

The final part of the drama is getting the hay home and dry. The bales thud out of the back end of the baling machine, like eggs from a large rectangular queen bee. The baler may have a sort of portable corral dragging behind it, which collects the bales as they drop. Every ten bales or so the corral is lifted, and the bales are left in a tumbled heap. The lorry or flatbed trailer arrives next and the bales are neatly stacked across it, in an interlocking pattern to keep them stable when they get above three or so deep.

Or you might stack the bales in the field first, especially

---

* More on silage some other time.

if the weather looks a bit wet. A good tight bale is surprisingly waterproof against the odd shower; stacking the bales helps reduce the exposed surfaces.

The final leg, from field to barn, is a delight. The phrase 'all is safely gathered in' keeps coming to mind as the load rolls home.

The stacking is a sweaty old business, but a labour of love, especially if you remember in advance to tape elastoplast round your inner knuckles before you start. That way you don't get blisters; quite so quickly.

If you have any sense of rhythm and balance, you soon get the hang of hefting bales about with the minimum of effort. With a hup and a bounce and a swing of the hip, the bale shifts from ground to shoulder height, via a bounce off the knee with scarcely any more effort than it takes to lift it off the ground.

Three people make light work. One unloads from the trailer; a second shifts the bales from trailer to the inner stack, then chucks them aloft; the third man has the hardest job, of clambering about on the growing stack, inter-locking each new bale into place, climbing ever higher and higher, right up to where the great big bristly spiders wheeze and roar under the slates and rafters, waiting to scrabble into your hair and whack you round the face with their nasty big hairy tendrils.

If you're careful, it's a fun job, testing your own dexterity and efficiency against the speed of delivery. Some middle-men deliver bales two at a time, or even three; but I personally regard that as cheating. They, on the other hand, tend to regard it as a Good Joke.

If you lose concentration, or work sloppily, you can hurt yourself by either falling off the stack (it's the embarrassment that hurts, unless your fall is broken by, say, the baling machine or disc harrow), or falling down a trap … a gap you've clumsily left between two bales. Nobody's clumsy enough to leave a full man-size trap, but you might twist an ankle or rick your back.*

Once it's all fixed, and the last few sweepings are tidied up, that pint of beer don't half go down well. So does the next one.

Smallholders are usually short of pastureland and have to buy hay in. If you're lucky, you can help a farmer with his own harvest and take some bales home as payment. Organic hay is rarer than the chemically boosted variety, and thus usually needs shifting further. If you don't own a truck of some sort, shifting hay can be a bit of a nightmare.

Our hundred and fifty bales, bought at £1 each, 'on the field', had to come ten miles. A haulier quoted an outrageous £50 for the job, so I tried elsewhere. One neighbour hadn't got a big trailer; a second had a trailer but no road tax on his tractor; a third had both his tractors in pieces. Various other ploys and hopes failed, including renting a Transit, which would have cost more than £50 and taken all day, ferrying small loads at a time.

So we gritted our teeth and paid up, as the barometer was dropping rapidly. And now our barn is full, and smells deliciously of … fresh hay. Sweeter than any perfume. It can rain as soon as it likes.

---

*Hence the term 'hay-rick'. Possibly.

# Hay and the other one ...

Before we came down here we did quite a lot of homework, but I personally was still a little uncertain about some of the more technical aspects of farming and growing: like the difference between hay and straw, for example. Anne never needed to find out, because she came from a then wildly rural area of Sussex and one of her earliest memories is of being five and driving an ex-army jeep round the trackless field the family house was in. Hay? Straw? Kid's stuff to her.

We had no jeeps in Liverpool; just big green trams and buses; and I wasn't allowed to drive them; not even a bit. It wasn't *fair*. (EXIT RIGHT, STAMPING SMALL CLARK'S SANDAL.)

Somehow I never got round to asking, and somehow we never discussed it, and somehow it finally ended up as being just too personally challenging to reveal that I'd got this far involved in our world-tilting scheme while carrying around this little sheaf of herbaceous ignorance. Somehow, it never occurred to me to look it up in one of our books. Odd, isn't it?

Never mind; I soon found out. Hay is cut and dried grass, suitable for fodder (an abbreviation for 'food fodder animals', I believe). Straw is what's left when wheat has been thrashed from the stalk. It is suitable for lying on, if you're a cow; voiding your copious bowel into, if you're a cow; and making dollies out of if you're a human being, keen on country traditions. These specific conditions are not cast in iron, however, so any cow with a free hour or two is free to make a corn dolly any time she likes.

**Cow Vanguards New Diversification Initiative: Minister To Speak To House
Sole Survivor of Witchcraft Purge Shocks Farming World with New Dolly Collection**

What we quickly discovered is that there ain't no straw in Wales. Nobody in his right mind grows wheat, which needs a long period of sunshine to ripen, which we can't rely on. You *might* get away with it, but the meteorological cards are stacked against you.

Occasionally people grow oats, however, as it/they tolerate/s wet better than wheat.

But realistically, all straw has to be hauled right across the country from the wheat and barley prairies in the East. They don't know what to do with it all, over there. They can't compost it for fertiliser, because they haven't got any dung. Guess why? Because all the dung-producing animals are in the grasslands of the West. This extreme specialisation is considered to be a sensible way of farming. Perhaps, just perhaps, we'll one day see the wisdom of mixing animals and cereals a bit more. Then we could use the straw to bed the beasts, and the muck to rot with the straw to produce organic fertiliser. Revolutionary suggestion.

Every autumn big lorries with big trailers wheeze and grind their way up the West Welsh hills, stacked high with wheat and barley straw, shedding occasional bales as overhanging branches take their toll. One is advised not to get stuck behind one of these monsters on the A484.

The major difference between the two types is that barley straw can be used as feed. It's not as good as hay, obviously, as the plants' nutrients have largely gone into the seed, but it makes a change and provides bulk for all those cavernous and fulminating rumens. We sometimes use it as bedding,* and watch with interest how much April will eat

---

*Yes, for the cow.

before despoiling it in her own inimitable and casual manner. No forethought, you see. None at all. No: 'Oh, perhaps if I stood over *there* to unleash the sphincter, then I could have this nice clean patch for breakfast or to make a new dolly with tomorrow.' Not a chance. Just pause chewing for a moment; stare harder at the wall; possibly hunch the lower back while gaping the jaw a little; then: 'We have lift off!', pause, and resume chewing.

I pause here to ask any Scottish reader whether the undoubted magic of porridge is simply down to its nutritional value: or does the filling effect come into it? I think our cows would help with the answer.

So ... I can now quite confidently distinguish hay from straw, although I've come across some 'hay' of such poor quality that the boundary blurs.

\* \* \*

One of the great romantic and mathematical delights of the World of Hay is 'how does the knotting machine work?'

Most people who look at a bale of hay think, 'Oh look ... a bale of straw,' and wander off to look at something else. The more philosophical of us are moved to wonder at how the baler twine, which enters the machine on a large roll, is somehow wrapped round four of the bale's six sides, *then tied in a knot, trimmed off, and set up to do the whole extraordinary business again and again.* What is more, there are two of these machines working in co-ordinated parallel. I find this an amazing triumph of engineering, but unfortunately I've never found the time to find out how the magic is worked. One day.

Polypropylene baler twine is one of those wonder materials that no farmer is ever without half a shedful of. It's used for tying everything and anything. Dogs get hauled round on it, errant tractor bits are tied back with it; fences are fixed with it. You tie your jacket round with it,* and will occasionally lace your boots with it. If you split it in half it's ideal for tying up sacks of spuds. And if you split it further still, it comes in for tying bunches of spring onions or radishes.

Naturally, your strings of onions and garlic are entwined round it, and the runners and toms and cues in the tunnels climb happily up it. You can plait it into very strong rope. I once saw a barn roof, after a gale, held down by it. You can make hay-nets and marker lines and doormats from it. There are even local craft courses in 'Fascinating Things To Do With It'. My new Spring Collection of beach and evening wear is made entirely from it.

If baler twine were to magically dissolve overnight, large areas of West Wales would cease to function entirely, and require government intervention, with the immediate promise of tankerloads of Canadian baler twine as part of the economic regeneration package.

As we are short of space, we only ever grew our own hay once. It was just an acre's worth, more as a sort of challenge than anything else. We'd got hold of an old finger mower (a very apt name if you weren't careful) which ran off the Fergie's own power drive. It was a bit old and bent, but it sort of worked. I sharpened the blades and oiled everything in sight, then frightened myself to death by the violent chattering of its

---

*A friend once turned up with his jacket tied round with a bungee luggage strap. I cut him dead.

teeth when the drive cut in. It had a definite look of the late Middle Ages. Torquemada would have owned one, or, more likely, a full set of seven.

It was quite an awkward thing to adjust, but with the right attitude and gritted teeth, I managed to hack down a fair proportion of the designated area, with hardly a divot to spoil it. We left the grass where it lay to dry off for a few hours, then I tried turning it with an old 'Cock Pheasant' hay-turning machine we found abandoned in our top field when we arrived. The tyres were flat and some of the less important bits were probably missing, but it made a fair go of turning the hay. It wasn't worth the trouble though for such a small quantity, and we eventually passed it on to Ken who could make much better use of it. We did the next turning by hand, with a couple of big hay rakes. A pleasantly aromatic and timeless job if you haven't got 20 acres to do.

We kept turning it, and a day or so later we collected it in with the help of our work party friends (more on work parties another time). This was a simple, fun job. Spread bedsheet; haul up an armful of hay; dump into bedsheet. Repeat. When the bedsheet is full, gather the corners; tie in some kind of knot; swing over shoulder; stagger amusingly in circles and intricate ellipsoids for several seconds while a new centre of gravity is established; stumble to barn; collapse, briefly; gradually prise apart super-tightened knot with prong of pitchfork; chuck hay onto growing heap. Repeat until tea-break. Repeat.

The work went so quickly, even with the tea-breaks, that I wondered whether we might even have tried cutting it by hand. But, realistically, we had no time for such whimsies.

And anyway, a dozen people slashing away with scythes and sickles and billhooks and cleavers in a disorderly medieval frenzy would have been bound to end in bloodshed with tetanus all round, at least two ambulance calls, and probably a steep fine.

## Hippy Loonies Go Mad Again – Moon Blamed

As it was, the only problem we had was with the hay stack. It was just a loose heap, of roughly the same size as the first atomic pile in Chicago. It also proved to have similar thermal tendencies.

Grass cuttings get very hot, and may even spontaneously combust. Our hay was not absolutely dry. When all our friends had gone, we had our dinner, and were just set for a crashed-out evening peering into the snow on our little black and white portable when I was suddenly stricken by doubt. I went back to the barn and thrust my arm into the hay. Then pulled it out again very quickly. It was HOT. The options were to leave it and hope that the barn and everything in it would still be there in the morning, or turn the hay.

I get anxious at times like these. If I'm pre-warned of something, I just have to act on it. I'd feel so stupid if I didn't, and then, you know …

## Hippy Loony Burns His Own Barn Down:
## see Editorial 'Bloody Hippy Loonies', page14

So we spent the rest of the evening turfing hot grass around, laying it as thinly as we could to let the heat dissipate and to allow for further slow drying.

It was fine in the end. And excellent hay.

# Wind, Winter 1995

I've never seen the point of wind. Rain at a pinch; but wind? I'm not talking lambent zephyrs here, but the stuff that hurtles off the Atlantic, like a wringing wet Rottweiler looking for trouble.

When we lived in our Edwardian suburb, a windy night meant a gaudy leafdrift for the kids to chase the dog through in the morning, or, in extremis, a loose or broken tile. Heavens! That was a rough night, and no mistake. It would be the talk of The Lounge Bar for a week.

But out here, all manner of things can happen. We've had two hundred runner beans stripped of all leaves, polytunnel doors, 6ft by 3ft, whirled to the far side of the field (one we never found again) and courgette plants blown clean out of the ground. We even discovered that an old touring caravan we'd stuck in the bottom yard out of the way had had half its back wall blown out.

The commonest damage, not surprisingly, is to buildings. Most constructions round here, as in any more or less depressed agricultural area, reflect the Brute Force and Ignorance school of architecture. Anything will do, as long as it is quick and cheap. Especially cheap. Stone walls and slate are out; breezeblocks and galvanised tin are in; preferably secondhand. Nothing wrong with that in itself. A shelter is a shelter. But it's so easy to underestimate the power you are up against. I discovered this the hard way myself when I leaned four large curved asbestos sheets against the lee side of an already sheltered shed as a handy bivouac for the cow.*

---

*I know. We ruin her, I know.

The heavy sheets were interlocked, and stabilised with breezeblocks, some on the top, and some at the base. One night we had a bit of a blow and the whole lot collapsed and smashed. No damage apart from the mess. Daisy knew a death-trap when she saw one and preferred standing with her head in a hedge in the time-honoured manner, until the storm had passed.

Galvanised tin, however, is dangerous. Unless it is properly maintained it eventually rusts round the nail holes and is ripped off by a gale and hurled through the farm like a poltergeist's dream. It's a brave man who goes out at night in a storm if he's got a roof that rattles.

One local man had no problems of that order with his new lean-to shed. He'd had it built on the cheap by a neighbour of erratic abilities known far and wide as Danny the Thief. All the tin sheets were very firmly nailed down, as per very firm instructions. Lovely job. Unfortunately, they overlapped from the bottom up, not from the top down. I don't know what the owner had stored in there, but it would have been nice and clean in the morning after the rain. I don't think Danny got the contract for the loft conversion.

Surprisingly enough, polytunnels stand up to gales very well, especially if you take the doors off. And the tighter the skin, the longer it survives, unless there's flying tin around.

A half-decent gale in autumn invariably means flooding of some sort. Leaves block ditches and drains, and water overflows. We will often have three inches of water streaming through the yard, pitting the tarmac as it goes, because a critical pipe has got stuffed up with sycamore leaves. Should we cut down the trees? Or hand pick the leaves *just* before the gale? Things to consider. Or should we just put a bigger pipe in?

There's not an outbuilding that escapes. Rain is blown under doors, under eaves, and, on one memorable occasion, clean through a breezeblock wall. The blocks were unpainted, and years of sou'westerlies had forced a way through the binding agent. Two coats of blackstuff solved that one. Unless lean-tos are solidly keyed in with lead, rather than patched on with hi-tec flashing, they too will leak. Guaranteed.

It's normal for animals to be missing after a brisk overnight gale. The cows are usually easy to find. They will have ambled to the least draughty place, usually in the farthest corner somewhere, turned their heads away from the blast, and switched off for the duration, working on the splendid philosophy they share with the Ancien Regime, that if they can't *see* it happening, then it *isn't* happening. All you have to do is find them.

Sheep, on the other hand, tend to burrow further and further into whatever shelter they can find. If you've got a brambly patch, you're in for a busy morning of hacking, clipping, and low blasphemy.

And then there's the powercuts. Three or four every winter, usually just for the morning, but sometimes all day or longer. Candles are essential, along with a rechargeable torch the size of a microwave. With the telly off, the stereo u/s, and the toaster out of commission, you suddenly realise how close to the Stone Age we live. A morning, or even a day without these things is OK. Quite romantic, even, as you sit round the piano, wishing someone could play it. But a week with no washing machine or kettle or central heating pump? ('No 'Archers'?' Anne chips in. I'm tempted to contribute something on the lines of 'It's an ill wind … ' but catch her eye and decide not to.)

No, I don't like wind. Still don't see the point of it.

### _Whistling in the Wind_

_Call this a wind?_
_While all the slates remain_
_while the grass is still rooted_
_while the sheep are still tucked under the cypress_
_and the cow's arse pokes out from her head-first_
_hidey-hole_
_in the shrubbery..._
_Call this a wind?_
_Ha!_

_I've seen better winds on a millpond;_
_a crowded summer beach;_
_in a field full of picnics..._
_storming away in a teapot._

_But it's blowing alright,_
_and twice as hard in the dark._
_Things skitter and clash in the yard;_
_things rattle that shouldn't;_
_and draughts eddy in the snug and battened sittingroom._

_The power's down;_
_the telly's off. No music_
_now._
_Two flittering candles between us two_
_and the elemental howl._

_Time to whistle, I think._

We had a townie visitor once who came out onto the field with me one soggy morning. He stood face on into the tail end of an overnight gale that had just flattened 100 feet of broad beans which would now need an hour's extra work staking up with bamboos and baler twine. He smiled, took a theatrically deep breath, and then told me how much he *loved* the wind. His smile faltered slightly and he looked away in some confusion as he caught the look on my face.

I was sorry about it. I didn't do it on purpose. Just ... you know: 'wind'.

# On battering butter

Hay, like grass, turns magically into milk.* Lots of milk. When Daisy arrived she was churning out nearly four gallons a day. What on earth were we going to do with it all? Pushing it through a pig is the traditional way of converting excess milk into storable food, but we didn't want to take on too many new projects at once. Cow now; pig later.

Our temporary solution was not ideal but not completely bad either. We made lots of butter.

If you fancy trying this yourself, this is what you do (unless, of course, you know of a better way, in which case skip the rest of this passage and move on to the next article, which is about the tractor. Or go and feed the cat or burn down a library. It really is up to you):

As soon as the stainless steel bucket enters the kitchen, pour the warm foaming milk into the biggest shallowest plastic bowl you can fit into the fridge. Leave it there till the next bucket arrives.

When this event occurs (NB: the cow may be working to a different timescale and/or agenda from your own; but it must arrive eventually), then the processing begins.

A few hours in the cooler will have caused the cream to rise sufficiently to be lifted off the milk. Jersey cream rises very quickly because the fat globules are larger than in other breeds' milk, and are therefore less homogenised. Other milks take longer, and can't match the Jersey's five per cent butterfat content either. Overall it's easy to see why the doe-eyed beauty is such a good beast for a smallholder. And they're not flighty

---

* Even more magically, because all the grassy juices have gone. Now the alchemy derives from just dry grass and plain water. Amazing.

or tremendously stupid, like some. Not vain, either. Just plain lovely. Some are excellent dancers.

The next bit is the fun bit. Scoop the cream off with a big colander spoon. It is so thick it wrinkles, and should peel off as neatly as a toupee off a pearl diver.

Store the cream in another bowl, again in the fridge. Your immediate problem now is 'What do I do with nearly two gallons of more-or-less skimmed milk, given a family of six, plus small dumpy dog and two semi-feral cats?'

We found our own solution. After taking off as much as we could sensibly use to cook with or drink, we gave some to the chickens, mixed with their meal, a saucerful each to the two rabbit-gorged and wild-eyed grimalkins, as much to Porky as she would groaningly accept, and the rest just went into the compost heap. Criminal, I know.

Yes, I know. We were better organised the next year.

Back to butter:

Every two days or so, wheel out the Kenwood and pour all the collected cream into it (I mean 'into the bowl', as you well know). There should be about three pints by now, unless you are inordinately fond of strawberries or heart attacks, in which case there will be rather less.

Click the balloon whisk in, and start whipping. After about ten minutes the cream will show flecks of yellow, and a thin buttermilk liquid will begin to separate. The colder the cream, the longer this process will take, so it's as well to let the cream rise to room temperature before whipping.

The trick now is to stop whipping before the near-butter winds itself up into an attitude and begins slopping the buttermilk over the edge and down your trousers. Slow the beater down.

Sooner or later, you will drain off the buttermilk and try to engage the dog's interest.

But the dog is by now virtually globular in shape after several weeks of soaking up your excess milky fluids, and in a state of near exhaustion after all those countless trips to the back door and slightly beyond. Pour it away if you can't face drinking it yourself, and give the proto-butter a good washing under the tap. It won't all wash away, believe me. Be bold.

Next, engage the dough hook and slowly stir the butter round to conglomerate it and squeeze out as much water as possible. Repeat the washing if you wish. The cleaner you can get it at this stage, the better; pure butter-fat is less likely to go rancid.* Then add salt to taste; or chopped herbs, if you're very fond of herby butter; or minced bloaters; or grated Viagra; or whatever takes your fancy, provided that you realise you'll be making a pound and a half at a time and will have to make the time to cope with the results of your actions. I'm thinking particularly of the Viagra option here.

My own favourite additive, which I never actually tried out because Anne wouldn't let me, would be a jar of chunky Oxford marmalade so as to save valuable time on preparing the morning toast.

You now thud your pound and a half (almost exactly slightly less than a kilo) of bright yellow butter onto a suitable surface, and bash it about, pretty much as violently as you

---

*Some people still like to make butter in one of the more traditional methods, by which you allow the cream to sour off, then lard it with plenty of salt. I've experienced this product, and was amazed at the flavour. It is *utterly* disgusting. Really, really horrible. Rancid blubber boiled in frog droppings comes to mind. I seriously recommend not going down this path.

like, for as long as you fancy. Scotch hands are the traditional tools for the job, but you can use slotted spoons, or a pair of old slippers, or whatever catches your eye. What you're doing here is knocking out any residual liquor, and making a useful storable shape. Have greaseproof paper to hand.

My first attempt at woodwork was to make Anne a pair of Scotch hands from a length of reclaimed pallet wood. I used a saw, a couple of chisels, a ruler and pencil, and one of those neat Stanley-type knives with snap-off blades for the final grooved channels. The resultant implements were a bit eccentric and wiggly, but worked well. Given a century or so of regular use they would no doubt acquire a splendidly rich patina and lots of admiration for the honest humble rustic workmanship, and command a two-figure price on *Antiques Roadshow 3000*. It's a pity we accidentally burned one of them. One Scotch hand is about as much use as a scissor or a yo.

\* \* \*

If you carry on making butter at the rate above, you will be building up a stock at the rate of a couple of kilos a week. By the time the cow's milk begins to dry off, after ten weeks or so, you'll have enough in the freezer for several months. Shame about all that buttermilk.

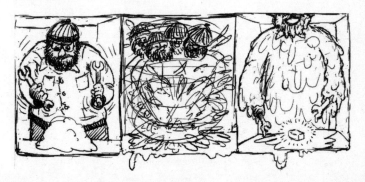

Pure butter straight from the freezer has all the spreadability of a breezeblock. Anne made some very good 'I Can't Believe It's Not a Breezeblock' spread by stirring in some sunflower oil at the dough hook stage, proportions arrived at by trial and error. I guess one could try other oils as well. Why not? Sesame? Lavender?

# The Tractor Ritual, Spring 1996

Our old Grey Fergie tractor gets a little light exercise every spring-time, but then it sits idle in its shed for the next 11½ months. It gets stiff, poor thing. This year it got stiffer than usual.

For a start, it spent all winter outside. This was not a delib-erate act of mechanical cruelty on our part, but an absurd acci-dent. For perfectly sensible reasons that I won't go into here,* a couple of friends spent some hours helping me to replace a perfectly good rear wheel and tyre with a terminally rust-riddled and punctured one. It took us the whole of a drizzly November morning, as tractor wheels are heavy and awkward things to shift around, and impossible to lift cleanly onto the studs of the wheel unless you are Jeff Capes and have x-ray eyes in your kneecaps. You need at least one assistant, and all manner of levers and shims and advanced shouting techniques.

As soon as we'd finally got the busted wheel firmly bolted on, it started raining, with that Old Testament intensity that everyone recognises as the first splash of winter.

So my friends slackened off their trusses, waved cheerily and went home for the duration. Obviously I couldn't garage the tractor with a busted wheel, so I chucked a tarpaulin over it, tied it down with baler twine, and hoped for the best.

In March, I tried to start it. Naturally, it refused. I tried my usual limited repertoire of cleaning plugs, waggling leads, polishing points and filling the battery with fresh custard. No good. Eventually a neighbour helped out with a huge battery and some industrial jump leads that would have sent Lazarus clean through the roof. Fergie wheezed back to life.

---

*Details will follow, one day … and again, names will be named.

Then it wouldn't go into gear. It seems that all the lubricant had drained away from the mechanism. We unbolted the top-plate of the gearbox and lifted out the stick. From what little we could see of the gubbins, nothing appeared to be broken, so with a four foot crowbar David forced a selector into first gear. 'Once the tractor is rolling, oil will splash over the selector unit, and all will be well again,' he said.

But of course we *couldn't* get it rolling because of the duff rear wheel which was so rusty it would simply have buckled and split and keeled the tractor unhelpfully over into a 20° list, suitable for barnacle-scraping perhaps, but not for light agricultural duties.

So David jacked the said wheel clear of the ground and fired her up again. Everything shook and wobbled horribly on the bottle jack, but it did the trick. Except it still wouldn't go into reverse. This meant it was pointless to even consider garaging it, as I'd never get it out again.

But patience and the crowbar prevailed, and we eventually forced it to acknowledge 'reverse'. We even replaced the good wheel. Fergie went back into its shed, ready for spring.

Then, one vivacious April morning, just as spring was teetering on the brink of springing … it snowed, and the shed collapsed on top of poor old Fergie. My friends of the previous autumn spent four generous hours of applied medieval engineering resuscitating the rusty tin roof with ropes and pulleys and doughty boughs; well, enough of it to get the tractor out, anyway. We slapped our thighs and quaffed mead in celebration. Then spat on the floor and belched amusingly for a while.

The final push came yesterday when I actually needed Fergie for work. Of course it wouldn't start, despite re-charging and boosting with a big battery and two sets of nasty puny little jump leads.* So I went through the starting procedure again. My ministrations had no effect. In fact within five minutes I'd snapped the top off a plug, and shattered the bakelite rotor arm by not replacing it full-square and true in the distributor after scraping the contact shiny with my penknife. I found a spare plug and prised another rotor arm out of the rusty scrapper we keep in a large convenient bramble patch; then recharged the batteries and tried again. God knows why, but it started.

So we set to ridging. Unfortunately, the soil was still a bit wet, and the tread on one tyre is, well, more of a ripple than a tread, and the tractor slewed and crabbed and pitched its way like a Mark 1 tank up and over the potato rows. A bit of a mess, but at least no mass attack by hordes of field-grey infantry. Then it hiccuped and thrubbed, popped, popped again, and conked out. Right in the middle of the field, as is traditional.

I took a deep breath and tried the Special Emergency Tractor Ritual which involves the usual plugs, points, leads, and battery procedures, PLUS a hefty kick to the rear tyre. From previous experience, I knew this to be virtually foolproof.

Nothing. I rang Alun the Mechanic, who dropped everything and rushed round.

'I don't understand it, Alun,' I said, handing over my tool bag to The Master. 'Is there any hope?'

'Just give it a try,' he said, still with his hands in his pockets.

It started first pull. That's the second time they've pulled that stunt on me.

---

*The Lazarus Leads were unavailable. A teacher friend who gets remarkably high pass rates at GCSE had borrowed them.

# Forget your Ferrari ...

It is a truth universally acknowledged that The Little Grey Fergie is one of the world's great masterpieces of design. Everybody who's ever owned one looks back on it with nostalgia. Our postman, when I told him we'd just got a Fergie, insisted on going to have a look at it and spent the next half hour patting its bonnet and telling me about the one he used to have. 'I wish I'd still got it,' he said, 'I'd put it on the mantelpiece.'

What is it that makes it such an icon? Several things. Perhaps the most important is that it did for tractors what the Beetle or Mini did for cars. It made them affordable. Before the Fergie, tractors had been great heavy lumping things that showed their derivation from the steam leviathans only too clearly. They were brute hauliers, basic and crude; and notoriously expensive to boot.

Harry Ferguson changed all that by designing from scratch in the 1930s and '40s. What did a farmer want? He wanted something well-built and reliable which was small enough to manage on his own, that would plough and harrow and all that sort of thing, and which would enable him to carry his implements from field to field without having to unhitch them and load them onto a trailer. He also wanted to be able to use his tractor as a mobile power plant which would enable him to use a winch or a sawbench in a far corner of a distant field.

This was quite a challenge, and Harry met it beautifully. The most noticeable thing to a stranger is that the Fergie is built on a human scale. Despite what I've just said about changing the rear wheel, it is not completely beyond the strength and ingenuity of one man to do the job. But take a look at a big modern tractor for comparison. If one of those huge rear tyres springs a leak, it is no trivial matter getting it repaired. You're unlikely to get much joy tackling it yourself with a couple of spoon handles on the front lawn. It won't fit on your front lawn for a start.

In fact, the Fergie was conceived and built on the scale of what it was due to replace: the horse. You really do want to pat it, the way you might a favourite cob. You wouldn't want to pat a huge John Deere from the prairies.

So the new machine was on the human scale at last. But its major triumph lay in its engineering novelty and attention to detail. Don't you find yourself smiling in disbelief at a machine that came supplied with a proper starting handle and toolkit, including a dinky little grease gun, and the minimum of different nut sizes? Isn't that just terrific? And look at the quality of the fittings. The brass where brass was needed; the high quality of the steel; even the nuts and studs inside the brake hubs are top notch. Quality everywhere. I loosened nuts that hadn't been touched for 30 years. They came off as smoothly as you could wish for. And went back on again with firm finger pressure.

We have two ridgers. One is a fairly modern Fordson model; the other is an original Fergie design. Both do the same job, and both fit on the Fergie tractor. But the Fordson is a coarse and heavy lump, where the Fergie one is elegant and light. The bars are perforated to save weight. The whole design ensures that one person can attach this tool to the tractor without risk of rupture or prolapse. It's as much a gem as the rest of the Fergie kit. Of course, the Fordson model is designed for deeper, heavier work, but it is still a dull brute by comparison.

The best is yet to come. How does the farmer carry his plough from one field to the next, without furrowing and ridging every farm track and council road between them?

Every tractor these days carries Harry's solution: the three-point linkage. This most elegant of designs enables you to fix your plough or harrow to three points on the back of the tractor. Thus you have a triangulated stability. These three points all rise up from ground level by a foot or so, using hydraulic power. This revolutionary and ingenious mechanism means every farmer is now truly mobile and can shift from job to job much more efficiently.

A further refinement is that the tool at the back may also be independently powered via the rotating pto (power-take-off) shaft. These days you can run everything from a hedge trimmer to a concrete mixer from your pto, and use it anywhere you can drive to: all thanks to Harry Ferguson's hydraulic linkage. Three of the little beauties, suitably adapted, even accompanied Edmund Hillary on his Antarctic expedition in the 1950s.

We used our little link box to cart fence posts and rolls of pigmesh around, and to haul firewood back from the brow of the hill nearest the margin of the wood. Once or twice we used it to fetch sacks and tubs of veg back off the field. We didn't do this too often, though, partly because it used carbon fuel and thus caused pollution, and partly because it used *expensive* carbon fuel: petrol.

A detail we hadn't thought through when we plumped for the TVO model was precisely how we would be using the tractor. Farmers use their tractor pretty well every day, for pretty well every job. If they're not ploughing or harrowing, they're towing a trailerful of stuff or beasts around, or hauling a cow or a visitor, or both (some visitors are highly eccentric) out of a ditch or slurry pit. Or they might be using the machine's awesome pto power to spread fertiliser or cut thistles or wallop fence posts in, or pump out a pond.

We, on the other hand, really only needed the tractor's high and sustained power for cultivation and muck-spreading; and that only happened once a year. Most of our other jobs didn't need all that power. And it really wasn't worth the fag of going to fetch the tractor and going through all the rigmarole of starting it up just to shift a dozen sacks of carrots a couple of hundred yards. By the time you'd got the tractor out you could have barrowed the sacks.

What has this all got to do with petrol? I hear you ask. Well, the TVO model has two fuel tanks. The single-gallon one holds petrol, and the seven-gallon one holds TVO/fuel oil/rendered suet. You start the engine on the more volatile petrol, then when it has properly warmed up (watch the little thermometer on the dashboard) you lean forward and twist the beautifully crafted brass butterfly lever which turns off the petrol supply and turns on the TVO. Lovely job.

It would take several minutes, especially in winter, for the engine to warm through enough for it to work with TVO. And most of our jobs only needed the motor to be running for a minute or two. We simply never got a chance to economise by using the nice cheap TVO! And because a tractor is not the world's most lean burn machine, we were forever nipping up to the garage with our jerry can for one gallon of three-star.

We did make the effort to economise, though. We had a 45 gallon oil drum and thought we'd get the local oil baron to fill it with heating oil for us. Not that easy; they only sell it by the 100 gallons. So we bought in two more drums. The tanker came and filled them for us, and yes, it was a lot cheaper than buying kerosene by the gallon from the garage. So the 100 gallons sat in the yard, just waiting: but we never quite got round to using it, for various reasons. Most of it is still there.

The only other problem with the TVO is the fag of having to remember to switch the brass butterfly back to 'petrol', approximately two minutes before you are going to switch the motor off. If you forget to do this, or stall the motor for some reason, you have to go through the Draining of the Carburettor Ritual. This involves holding the plastic cap off a spray can, or something small and similar, under the carb while you turn a little lever that drains the TVO out, partly into the cup-cap, but mainly all over your fingers, and down your trousers if it's a bit windy. Two cups full is enough, assuming you *have* actually remembered to turn the butterfly back to 'petrol'. Otherwise, you will wonder why the damn thing still refuses to start, despite hauling on the starter till the battery fades. Usually it is the rank stink of paraffin from the flooded carb that reminds you of your oversight. So you switch the butterfly firmly to 'petrol' and start the draining ritual again; then try the starter; then clamber off the seat, rummage in the tool box for the terminal spanner (great name), unhitch the battery, and slog off across the field to the shed where you can begin re-charging the battery. Again.

\* \* \*

I asked Alun if duff vehicles often sprang to life merely because he told them to. 'You'd be surprised,' he said.

Not totally surprised perhaps: it's often struck me how quickly an empty house suddenly appears derelict. It develops a forlorn and vacant look, and bits of guttering seem to drop off much more quickly than you'd expect. Am I alone in noticing this? I think not.

Dai's old lorry was a splendid example. It was stylishly finished in orange and green and had served him faithfully for 100,000 miles. He had renovated half of it, and replaced the other half, but it eventually would have to go. His hand was forced when a traffic cop pulled him over in Manchester for being clearly overloaded. 'Let's have a look at your tachograph, son.'

'Haven't got one. No. Sorry.'

(PAUSE)

'MoT then?'

(PAUSE)

'Haven't had one since 1981.'

(PAUSE WHILE 'THE BOOK' IS WAITING TO BE THROWN. THE PC CHECKS, POKES AND PRODS EVERYTHING POSSIBLE.)

'Tell you what, son. You've looked after this old heap well enough, even if you haven't got a ticket. We've got things in the lorry pound over there with full MoTs that are death traps. But this old wagon's really got to go, and you know it. Just take it straight home and take it off the road immediately. Fair do's?'

'Fair do's. Thank you.'

'Mind how you go. And I mean that most sincerely.'

Dai got his wagon home, still running well. He took it off the road immediately. Within the month a bigger lorry came to tow it away to the scrapyard. Dai discovered an unprecedented pool of oil under his old truck. And suddenly the wipers had ceased to work; and as they began winching it onto the breakdown vehicle, a side light unit fell off. Very odd.

# Rabbits, Summer 1996

The first rabbit I shot was wearing a bra. Hard to believe, but true. We'd brought the kids' two pet bunnies with us from suburbia and kept them in a cage, away from the murderous teeth of the fox. One spring day, we thought the bunnies would like a run on the lawn. But surely they would abscond, wouldn't they? Whiff of Freedom? Call of the Wild? Thundering Across the Serengeti, Ears Flopping Heroically in the Dazzling Sunlight? Who could resist? Best think in terms of moderate restraint.

We still had the harness we'd used when Porky was a puppy, and with a bit of a fiddle it fitted Blackie (he was black) a treat. One rabbit with a token of freedom.

Five yards away, we were planting fruit bushes. Suddenly there was an appalling scream. Twinkle-ears, our tortoise-shell cat, had got Blackie. We beat her off with boots and a spade, but Blackie was still screaming. There was no blood, so I suppose the cat must have broken his back. I should have given him a quick rabbit punch to end his misery. I know that now, but at that moment of truth, when I was confronted with the necessity of decision, I confused respect for life with moral cowardice. And, to be fair, for the first minute or so I thought he might just be plain terrified and would soon calm down again. But he didn't.

I carefully carried the screeching beast to the Woodbarn (why the Woodbarn? I've no idea ...) while Anne fetched Paddy's air rifle. I had to shoot the poor thing. I knew that. It seemed to take ages. I kept realising that that slight press on the trigger would take a life. Eventually I did it. I'll never forget the scarlet on the black and the kicking. Then the mysterious stillness. Then I went to pieces. My first time, as they say.*

Since then, I've shot a hundred rabbits. I still don't like it. But if five rabbits eat a sheepsworth of grass, on our tiny acreage (and even tinier hectareage) we can only keep sheep at all if we keep the rabbits down. We tried expensive electric anti-rabbit netting to keep them off the veg. It didn't work very well. One morning we found a full-grown rabbit, eighteen inches long, caught by the hips in the mesh. Presumably electrocuted. And presumably any slightly smaller rabbit could Fosbury Flopsy in and out at leisure.

Some of them bit their way through the unpowered vertical strands, presumably in order to make the mesh big enough to jump through. How did they know? Trial and error? Logic?? 'Owww!!! By jiminy, that really hurt! Better not try that again. Hang on, though … let's think about this … it was the horizontal strand that hurt. Perhaps the *vertical* one will not hurt? Ah! See! No tell-tale thread of stainless steel conductor! Let's just try it and … bingo! …'

I believe there is a school of theoretical behavioural psychologists who say animals have no intelligence because they have no mind. I wonder if they've ever actually seen any real animals, rather than the miserable things they multiply in cages?

Eventually we realised that, yet again, lo-tec was the answer; this time in the form of two foot high narrow-gauge chicken mesh, with the bottom six inches angled out and buried, to stop the varmints digging under it. It took days to

---

* (ENTER CHILD, FRESH FROM PLAYSCHOOL.) 'Where's Blackie, Mummy? I got a dandy-lion for him.' 'Daddy shot Blackie not half an hour ago, love. Would you like some milk?' 'Waaaaaahhhh. Hobbible Daddy. Waaahhhhhh …'

fit this outer sheathing of mesh to the 1600ft perimeter of pigmesh, wiring it on as we went so that sheep and calves couldn't push it down again when they stuck their hooters through the pigmesh to try and plunder the crops within.

But it was worth the cost and effort because it did actually keep the veg safe at last; at least until one beast got in, God knows how.

Needless to say, it was a pregnant female who did her bit for rabbitdom with never a thought for incest or Leviticus. We had a round-up one day and chased them all out with the wildly enthusiastic help of Porky. Rabbits can jump three feet high, incidentally, if the alternative is a terrier's undivided attention. Not a lot of people know that.

But we can't realistically fence the entire perimeter of the farm, or even just the pasture, with angled-out chicken netting. We've thought long and hard about it, and decided that the only way to keep rabbits off the pasture, apart from reducing the cover, is shooting. At least, that's our own considered view. Ferreting works, but it takes more time (and ferrets) than we've got, and it's a terrorist tactic anyway. 'Sport', some people call it. Snaring is horrible. In a fit of despair I did try snaring once. I found the little giveaway runs in the grass –

> 'Rabbit him go this way, Kemo Sabe.'
> 'How can you tell that, Tonto? A twisted blade of grass? A lingering aroma of some kind?'
> 'No. Him just run over my foot.'

– that tracked back to particular portals in the perimeter pigmesh and carefully laid the brass wire loop in the time-honoured manner. But they must have been watching me through little furry binoculars, or whatever. Or perhaps they just smelled my fear of success.

Anyway, I was so relieved that I hadn't caught anything, that I took all the snares up again the following morning and never tried again.

Shooting is quick, usually painless, and 'fair', especially with a high-powered air rifle. I once shot a fleeing rabbit stone dead at nearly forty yards. Luck? How can it be luck, when it was exactly what I was aiming to do?*

Ironically, the cats catch more rabbits than I do. They do it their way. You don't really want to know how, do you?

We once rescued a little one they were tormenting, and tried to calm it before putting it in a cage next to its two glossy and domesticated cousins. But it didn't work. It wouldn't eat or drink and just huddled in the corner, and just plain died overnight. Perhaps we should have left him to the cats?

Foxes? There's a joke. As long as we have chickens and ducks, the rabbits thrive. Natural, I suppose. You don't see restaurants advertising Rabbit Kiev or Rabbit à l'Orange, do you? Foxes aren't stupid.

The buzzards seem to catch a few, judging by the little drifts of grey-brown fluff we occasionally find. But we've never caught one in the act. It must be a mighty accomplished buzzard that can swoop from a clear sky, undetected by those sensitive and panoramic eyes.

Over the last couple of years rabbit numbers have dropped considerably. Don't know why. Myxy?

Myxomatosis does still take an occasional toll. You can usually tell when it's in the district because you see rabbits

---

* By the same token, can a golfer's hole-in-one be called 'luck'? Tricky philosophical problem. I might try it on the postman tomorrow morning.

puffy-eyed and hunched and silent in places where they shouldn't be, like in the middle of the yard. I stood watching one once, until it suddenly sensed it was being observed, whereupon it leapt up and ran off at full tilt … straight into my leg, then into a wall, then into the 'safety' of the woodland.

It's a nasty thing, myxy, but it must be said that it relieves me of a hard necessity.

\* \* \*

I became reasonably proficient at rabbit shooting as I learned to cope, at least to some degree, with my revulsion and distress at killing things. I even learned a little of the ways of the hunter.

I learned to dress in muted colours (not that I was normally given to strident shell-suits or end-of-the-pier ensembles) and to be sure I wasn't wearing anything that might flap about distractingly in a gust of wind. I also learned that you should move slowly and pause often, keeping your head down. Think 'cow'.

Take care to lift your foot and place it carefully, rather than stride. Think 'grazing cow'.

Don't rustle or creak. Think 'arthritic cow'.

Keep the long snout of the gun out of sight. Think 'rabbits are smarter than they look'.

The rabbits come up from the woods, into the bottom paddock mainly. My only approach is through the orchard, from above them. Thus my head will be visible to them above their horizon, before I can see them at all. Slow … slow: look slowly up … if one catches your eye, hold your gaze. Sometimes they freeze, and only move if you look away. I once saw a rabbit staring back at me from between two potato rows. I charged a visiting friend with holding the critter's eye while I went back to the house for the gun. As I returned, they were still staring each other out. 'Thanks,' I said, as I approached, gun loaded. 'No problem,' he responded, and looked across at me. A millisecond later the critter had disappeared. He never even saw the gun.

So … look carefully down the slope …

There's four of them within a yard of each other. Three nibbling, one washing its ears. That's the one. He's preoccupied, and less likely to see me slowly raise the gun to my shoulder. But one of the nibblers spots me; stares; freezes; thumps; turns tail; and all four of them hurtle back into the cwm before I can take aim.

That's a typical hunt. Perhaps one time in three I get the gun to my shoulder. Then it's a careful aim at the head, hold the breath, and squeeze the trigger. There's a strange PHWATT sound and one of the rabbits falls immediately flat. The others scatter. That's that.

They won't be back for at least twenty minutes. So, collect the victim and go for a cup of tea.

In practice, it's hardly ever worth going back again. They do return, but they're naturally very wary. I've only twice ever shot a second one on the same morning, and that was when there were so many we were tripping over them in the bedroom. Well, almost. They'd eaten the 'safe' grass down to the bare earth so they had to venture further and further from their bolthole to safety if they were to find anything to eat. The grim smallholder was waiting.

The best tactic is to tiptoe through the orchard as early in the morning as you can manage, and again at mid-dusk, keeping a weather eye on the gap in the slurry pit on the right, as they often infiltrate the orchard from there, intent on ringing the fruit trees and killing them. Pure spite, wouldn't you say?

One beautiful spring morning I made the trip, slowly treading my way through the glistening grass. It was one of those magnificent days full of birdsong and light. The very air seemed to be expanding by the minute.

As I was passing the gap to the slurry pit I glanced across. There was a little baby bunny, all curves and bright eyes, and little stubby ears; obviously on his first trip out alone. He was looking at me, quite fearless, and bright as a button. What a beautiful little creature. Then he sat back on his haunches and started to wash his ears with his paws. Pure Disney.

I was hunting rabbits not out of hate or vengeance, or for 'sport', but to try to retain enough grass for our growing flock of sheep. What should I do? I raised my gun and …

And what? Did I or didn't I? What would *you* have done? What did I do? Answer at the end of this item.

# One for the psychos ...

The traditional farm gun is the double-barrelled shotgun, and I tried one once. It was a neighbour's 12-bore which took a cartridge the size of a standard sausage and had a very big hole at the business end. Two very big holes, in fact, both of which pump out a rapidly expanding cone of small lead pellets, lethal at 80 yards and very irritating or more at 100 and upwards. You *can't* miss. It's the same technology Nelson used at Trafalgar, and he didn't miss either.

There's the rabbit, about 40 yards away. Quiet, now. Still. Slowly raise this whopping great blunderbuss of a thing to the shoulder. Hold firm. 'Discharge your fowling-piece when ready, Mr Hornblower' ... squeeeeze the trigger ...

The rabbit looks startled, which is sort of gratifying, then quietly hops out of the way. The second barrel misses as well, with an even bigger blam.

'Do you know what you're doing wrong?'

'Missing?'

'You're aiming, not pointing. You don't aim a shotgun, you point it.'

I found this quite confusing news. Surely if you hit something by pointing, you hit it better by aiming? Not so, it would seem.

The twin explosions had cleared the fields of everything larger than spiders for miles around and I had to wait for another day to try again, with much the same result, although I'm pretty sure I perforated the tip of one furry ear, judging by the way it dropped and twitched before rising again and following its owner into the brambles at very high speed.

Four shots: four misses. And four wasted cartridges at 50p or whatever a time. Hmm. Also large amount of lead pellet sprayed into the sward. Heavy metal not terrifically desirable for biological processes. Hmm. Also no depletion of rabbit. Also distinct rattling of at least two fillings, a swollen shoulder, and one eyeball that seemed to be idly rotating of its own accord.

Paddy's little target airgun wasn't much use either. It was OK for perforating single sheets of damp newspaper or knocking tin cans off posts at twenty feet, but frankly you don't usually get within twenty feet of a rabbit, and certainly not if it's sitting on a post. It wasn't very true, either. To be sure of hitting the can you had to aim several inches to the left and up a bit.

But an airgun seemed like a good bet in principle. In his army days, Dad had been a pretty good marksman* and he came home one day with a Webley Vulcan, a very serious airgun. Instead of the normal .177 pellets, it used .22s. This is the same calibre as a small rifle, and at close range it is a very effective tool. Tin cans dropped off posts and scuttled for cover at the very mention of the Vulcan.

It took a bit of getting used to, but I found it much better than the 12-bore. Much cheaper, too. Five hundred pellets cost no more than a couple of cartridges. And the much-reduced kick meant your teeth and tongue were safer. And

---

* Particularly with a three inch mortar, apparently; but they're hard to get hold of and you probably need a special licence, even for rabbiting.

there was far less lead sprayed about the place. And it was quieter; and satisfactorily lo-tec. Splendid. And if you hit a rabbit, it stayed hit. You had less chance of cruelly wounding. Or so I found. No doubt a hot-shot with a shotgun would say just the opposite.

A visiting friend distracted me down the shotgun path again, however. Rob had a smallholding near Nottingham which we used to visit before coming down here. He kept rabbits at bay with a 12-bore, and to keep the costs down, he used to re-load his own cartridges, using a pair of jeweller's scales, a flask of bird-shot and a horn of home-made black powder.

One day something went wrong, and a blowout in the chamber when he pulled the trigger nearly cost him his life, never mind his hearing. Since then he'd been more cautious, but still used a shotgun, albeit of a smaller bore, and always with store-bought shells.

This gun was a little Russian job, of .410 bore, and with just the one barrel. It did, however, have a magazine that took two extra shots (any more is illegal, it seems. Can't think why). Cartridges are cheaper than the 12-bore's, and you can buy short or long varieties. The long ones have greater range.

He shot a couple of rabbits with it, and took Paddy out with him, and showed him how to skin and gut his trophies. But Rob was going to be moving from his smallholding soon and wouldn't need his shotgun any more. Would I like it? Well, would I?

It was smaller and lighter than the 12-bore, and thus probably aimable, I thought, and might come in handy one day, even if only for lifting carrots on a frosty morning. So …
Yes. How much?

Then Rob wondered if I was going to be using my wah-wah pedal much? Well no, actually. I don't hardly play my electric guitar these days, and can't remember when I last creatively wah-wahed. Swop? Why not?

So the deal was done. One noisy and dangerously anti-social machismo-enhancing bit of kit, in exchange for a shot-gun. Fair deal.

But the little shotgun was no more use to me than the 12-bore. I still couldn't hit a thing* and soon reverted to the mighty Vulcan.

Even so, I had to go through the palaver of getting a licence. There wasn't any problem, although the police made a big fuss about keeping the gun under lock and key in a steel cabinet. A neighbour who was a gun-dealer told me this was all bluff. The cabinet is a recommendation, no more. It is not a legal requirement, and would certainly cost more than the gun, with or without a wah-wah pedal thrown in.

My view is that anybody who breaks in and sees a gun cabinet will definitely steal the gun, and probably the cabinet as well. On the other hand, if there is no cabinet, he is unlikely to ferret around under the great stack of grubby coats and waterproofs in the corner just on the off-chance of finding a small shotgun. A long-lost glass eye or roller skate, perhaps, but not a shotgun.

---

*I tell a lie. I did once hit a tree.

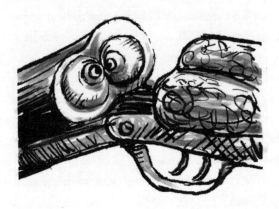

So, for several years I had a gun licence. It never ceases to amaze me that they issued one to me, as the photograph I submitted was taken in a Woolworth's booth one very rainy afternoon when I'd forgotten my hat. I'm staring at the lens, not best pleased, with rat-tailed hair and a look that can only be described as wringing-wet-murderous.

Perhaps all gun licence pictures look like that? 'Smile, Mr Capone! Say "cheese"! Oh *please* make the effort, Mr Droopy-Drawers!'

The time eventually came when we realised I was never going to use the gun again, so I asked the dealer if he'd buy it. He did, but for a pittance; and most of that was on account of the 30 cartridges that went with it. Moral: if you're going to get your money back on a shotgun, get a Purdy or a sawn-off.

What did we do with all those rabbits? We stewed a few, but they were such fiddly and time-consuming things to clean and chop, that we eventually didn't bother, and left them as a caution to the others and a light snack for the crows and buzzards. I tried to save a couple of pelts to make useful slippers or sensible underwear from, but they were very thin, and required more skill, time and attention than I had to offer.

\* \* \*

Another friend had a very posh air rifle indeed. It was a sleek German affair, narrow bored, with a telescopic sight the size of a cucumber, and was clearly designed for circumcising pigeons at a thousand metres. Jim thought it might be fun to shoot some of our surplus rabbits with his surgical howitzer, if I was willing.

'Certainly,' I said. 'There is no shortage.'

So Jim turned up with a Landrover full of straw bales, and built a cosy hide for himself from the gleaming blocks of straw, elegantly pierced and crenellated for maximum field of fire, just about on the brow of the hill. He opened his camp chair and his flask of coffee, and his fresh Bacon Lettuce and Tomato baguette, and sat down to wait, with his wallet of individually milled vanadium-tipped tungsten-niobium pellets readily to hand (I'm making this bit about the pellets up, as I expect you've guessed. But they *were* a lot posher than mine; in fact they would have looked rather good on a necklace and matching tiara).

He sat all morning and saw one rabbit that took one startled look, then beetled off through the bracken before he had a chance to pick up his gun, never mind adjust his high-resolution crosshairs, allowing for windspeed, humidity, and curvature of the earth.

Jim never came back again, but never said why. I think he thought I'd misled him when I said the place was heaving with rabbits and he'd be sure to bag at least four and a half brace.\*

---

\* 'Four and a half brace' equates with two and a quarter pairs of braces. Haberdashers take note.

I suspect, however, that the sudden appearance of a Siegfried Wall of brilliant straw bales on their normally grassily bland horizon may have given those cautious bunnies pause for thought. And then there was the whiff of coffee and the mixed perfumes of the BLT.

'Fancy a nibble?'

'Best not. There's a bloke up there with a gun.'

'Oh yeah. I can smell the tungsten-niobium, now you come to mention it.'

\* \* \*

We were going to demolish the straw bale parapet to mulch the apple trees with, but didn't want to pre-empt Jim's non-return, if you see what I mean. Those dozen shiny bales gradually faded and dulled, then sagged and sogged; and eventually they simply disappeared into the soil without trace, except for a dozen or so empty hoops of polypropylene baler twine. Ain't Nature wonderful?

\* \* \*

So that's rabbits and guns.

Did I shoot the little baby bunny?

Yes, I did. And hated myself.

# Bees, Autumn 1996

Bees. What are they to you? Gentle busy creatures of a social-istic bent, garnering nectar for your delight? Or brainless fascists, all spleen and venom?

I'm afraid I fall cleanly into the second camp. I've talked to several beekeepers about my experiences and without exception they have looked at me with an expression half-incredulous and half-pitying. It's as if they're waiting for me to deliver a punch line which will confirm that everything I've said has been some sort of elaborate joke, and that bees are really jolly nice little furry people.

Not so. Not my Dad's bees, at any rate. 'Vicious little bastards' is closer to the mark. They would fly straight at you, into your hair if possible, and harass and pursue you without mercy. I've lost count of how many friends have been stung. One, who was hypersensitive to bee venom, didn't dare visit for years.

Have you ever had a bee in your hair? They don't idly drone, as in meadows on a sleepy afternoon; they zizz and fizz like an amphetamine-powered dentist's drill; and they are very close to your ear; and you can't shake them off; and they're going to sting any second NOW! **OWW!** And even though you know swiping them off will rip out their sting mechanism and cause them a lingering death, you don't *care*. The BASTARD! Ow! And you hobble indoors, trying to remember whether it's vinegar or bicarb you need. Then, with a kamikaze zzizzz, zooom, and wallop, another one comes from nowhere. And the frantic pantomime starts all over again. **OWWW!!**

Surely I'm exaggerating? No, I am not. Ask my wife, chil-dren, or any of our ex-friends.

They were not constantly vicious; only sometimes. We never quite worked out what got into them, but on hot days you had to beware. Can you imagine easy living, fish jumping, cotton high; all that; temperature in the mid-eighties … and having to jog briskly from the back door to the veg patch in an anorak and balaclava? We can. We've done it. Regularly.

Once we reached the veg patch we could relax. Presumably the bees had something busier to do than patrol so far from the bunker.

But two sets of workmen had to abandon their work. One pair was attacked while slating a roof. They put up with it for twenty minutes, but both got stung and went off in a huff, muttering something about Pearl Harbour and danger money.

The other victim was John the WWOOFer, who hurriedly scrambled down from painting the top of the 20ft dutch barn with dead bees and bitumen in his hair.

Why were these bees so horrible? We still don't know. The hives were much closer to our house than to my parents', and as Dad only ever approached them in full spaceman fig, to either pinch the honey or add the sugar, he never really fully understood our gentle hints. (Politesse prevented us from revealing all.)

However, he did eventually buy in a new, allegedly docile, queen from Brother Adam's Hatchery and Nursery for Peaceable and Radiant Queens at Buckfast Abbey, but it made no difference. She may well have been peaceable and radiant in herself, but her offspring were as rabidly nasty and suicidal as ever.

A friend later pointed out that the hives were rather close to a high tension cable. I wonder?

Last year Dad decided the bees were getting to be too much work, and sold them. Sad to see his hobby go, but nice not to need a balaclava in July.

# Stick it

Dad had all the right gear for honey making, if 'making' is the word, because strictly speaking it's 'robbery', isn't it? 'Candy from a baby', sort of thing.

Sticky, runny, candy from squiggly, squirmy, throbby, *grubby* little babies, wriggling in identical regimented rows and columns, like ...*

When the time was ripe, in early autumn, Dad stuffed his smoker with corrugated cardboard, wisely ignoring my advice to pep it up a bit with a pint of petrol so as to get a decent jet of blue flame instead of just that feeble thread of smoke. He then donned his bright white bee-proof gear with contrasting black veil and trudged off to rob the hives, tugging his purpose-built home-made wooden trolley behind him to bring the booty home in. He looked like a spaceman who'd fallen on hard times.

The bees seemed not to mind too much being smoked down into their basements while Dad levered out the bulging frames, cracking them from the hangers with a special tool. I don't suppose the bees had too much option in the matter, come to think of it.

'Hey ... there's a bloke up there with a special tool, nicking all our winter food reserves for the little ones and orphans.'

'Best sting him to death, then.'

'No, like I say, he's got this special tool, and it's all smoke as well.'

'Oh, that's different. It must be God again.'

'How d'you mean?'

---

* I don't like the direction this is taking, so I think we should move on.

'Well, where else do you think all that terrific pure sugar food comes from? God takes out all that rubbishy home-made stuff of ours, and gives us back pure and refined sugar food instead. It's part of his plan.'

'Oh. What about all the smoke?'

'That is how he appears to us. Did you see a pillar of fire or flame, by any chance? Bluish, possibly?'

'No. Just …'

'… just the smoke? Oh dear. He must be angry with us again. Best keep your head down. He can be a bugger when he's in a strop.'

Back in the kitchen, Dad de-caps the slabs of honey cells with a quick sweep of a huge hot palette knife, one sweep for each side of each frame, then hangs the frames in the honey extractor. The sticky caps thus sliced off the tops of the honey cells are called 'cappings' and are wiped stickily off into a suitable bucket, tub, or vase for treatment later.

The extractor is a very impressive stainless steel centrifuge, the size of a large stand-free boiler. It takes four frames at a time, containing perhaps 12lb of honey in all.

The fun starts when you turn the handle. Slowly, the inner axle rotates; then a little faster as momentum builds up; finally the whole tub drums and thrubs and begins a little curving precessional dance round the kitchen floor, as the honey within is forced out by the centrifugal force and smacks into the inner steel wall, before draining glutinously into a deeply gooey mess at the bottom of the drum.

Mainly honey, but odd bits of wax and propolis and the odd dead bee or component part thereof mixed in. It doesn't look too promising at this stage.

The empty frames are levered stickily out, and the next four go in. Turn the handle. Twirl ... thrum ... whizz; then haul the empties out, again 'stickily'.

'Sticky' is the key word here. It is hard to overemphasise the stickiness of every aspect of this operation. You lay down newspaper on the floor, of course, several layers if you are wise (see Coco-the-Clown, below) and hermetically seal all doors and windows against outraged intruders of the apine persuasion who might attempt to distract you by stinging your hair or burrowing down your ears, thus causing you to thrash wildly round the kitchen, kicking over bucketsful of honey and empty frames and scattering gooey cappings and tools to the four winds, walls, and most of the ceiling.

You also act slowly and deliberately, taking every conceivable care to keep every trace of honey within its allotted bounds.

Oh, vanity.

The English language does not contain enough words to begin to express the range and quality of stickinesses involved here.

There is the ripping-stickiness of rubber slipper sole on vinyl flooring; the gummy-stickiness of the jumper elbow into the little pool that has drained off the handle end of a dislodged tool and onto the table; the Coco-the-Clown stickiness that gums the newspaper on the floor firmly to your shoes, so you blunder about with a two foot square snow-shoe on each foot; there is the super-glue stickiness that prevents you from ever properly letting go of a jug handle without carefully peeling your fingers off it, one by one; and there is that faint, almost ethereal or miasmic stickiness that eventually comes to lightly grace every surface throughout the house like a gentle blessing: first the table and tools, then the taps and towels, then the handles of cups, the edges of pockets and the arms of your spectacles; and finally, via the doorknobs, to the buttons on the television and the remote control, and ultimately to the ears of the dog, which drives him nuts as he rushes round in circles, trying to lick it off.

The good news is that he will clean up all door knobs, etc., if lifted up, held roughly horizontal, and carried round the house for long enough.

But that's another story.

So, yes … it's a sticky business.

Once all the frames have been whizzed and emptied, you move on to stage two: filtering.

You may possibly have foreseen that this is a very, yes, 'sticky' operation too.

In a good year, Dad's three hives might give well over 50 frames. Each one must be decapped with the hot knife, and those sticky cappings have to be put somewhere and each of the 50 or more emptied frames also has to be put somewhere, each still gently oozing. The potential for total mess is considerable already.

Now it moves up a notch.

You lift the centrifuge mechanism out of the steel drum and allow it to drain. This can take as little or as long as you like, as it is actually an infinite process. Never will the final drop of stickiness actually drip. At some point you have to give it a quick wipe with a knife or rag or shirt-tail, and then just abandon it, preferably in a sink or large bath. I would recommend in a jacuzzi at the bottom of the garden; at the bottom of somebody else's garden would be even better.

Next, very carefully drop the tapering colander-funnel

into the drum. It locates onto the top rim. Drape your filter inside the funnel.

You now have all the grubby honey at the bottom of the drum and the filter at the top. Thus you need to get the honey shifted back up to the top again.

The tap at the bottom of the drum allows the honey to flow into a suitable bucket for the transfer process. Note: this tap is surprisingly swift-running and totally silent. It is unwise to answer the phone or a call of Nature, no matter how urgent it may seem. Nothing short of Armageddon will compare to the trauma of the weeks of mindless scrubbing you will let yourself in for if that bucket overflows.

You filter the lot, taking paranoid care that you have turned the sticky tap OFF after the last of the grubby honey has drained out. Can you guess why?

So now you have maybe 100lb of honey or more, just waiting to be bottled.* Anyone who has ever bottled jam knows that Murphy's Law (Jams, Preserves, and Viscous Fluids) states that no matter how much meticulous care is taken in the process, *'between 0·01 per cent and 0·05 per cent of the total contents [by weight or volume, whichever is the most inconvenient] will find its way onto the outer surfaces of the jar, tin, bottle, or any other such container used for the bottling process, along with any lid, cap or other closure used for the sealing thereof'.* There's no way round it.

So you resign yourself to a certain amount of wiping and smearing and wiping again. Then it's done.

---

*You can't say 'jarred', can you? Boxed, bottled, tinned … but not 'jarred'. And some people think language is rational.

The sight of 100 pristine amber jars lined up on the kitchen table almost makes up for all those hours of hive preparation, boiling up sugar water, carting heavy jars of syrup downhill and heavy frames of honey uphill, getting stung now and again, fitting excluders and blocks, buying in all those jars and that Grand Panjandrum of a centrifuge … and all that stickiness, everywhere.

Or so Dad tells me. I can't see it, myself. It must be a labour of love.

Is it cost effective? No, not by commercial standards; but is must be a rewarding feeling to know that the honey on your morning toast is all your own work.* Our morning toast was completely home-produced; well largely anyway: we did buy in the flour.

Dad supplied the honey and baked the bread; Anne made the butter and baked bread as well. And, yes, it was unbeatable.

* * *

Dad sold a few jars and gave a lot away. He wasn't in it for the money.

The other by-products of bee-keeping are interesting. The cappings are of the finest whitest wax. After straining, you wash them and store them away from the predations of any passing wax moth, then use them for making fine polish. Some people make cosmetics. Dad made himself a sort of hand-cream from cappings and pure turps. He's a great bowls-player and uses the cream to give him a reliable grip on the bowl.

---

*I have not yet consulted the bees on this point.

Propolis is a sort of all-purpose bodging cement the bees use for stuffing cracks and general DIY. It too can be used in cosmetics, but it was just a nuisance to Dad, as it cemented too many things together when he was getting the frames out of the hive.

Royal jelly is useful, they say, for its astonishing rejuvenatory powers over skin and female parts and just about anything else; in my own dogmatic and prejudiced view it is suitable only for casting into moulds as royal jelly babies to chuck at the villain at the opera.

* * *

Dad once entered a couple of jars for the annual honey competition at his Bee Club. His two jars were immaculate, but he didn't win. He did come home rolling with laughter, however.

If you take pause for a moment, you can see that judging a pot of honey is not like judging a prize marrow or a bunch of carrots, or even a jar of marmalade. There is no skill in growing the unit-thing involved, and no recipe. All the bee-keeper does, as far as the jar of produce on the table is concerned, is open the tap, close it again, screw the top on and slap on a label. The award should come from the Institute of Packaging, not a bee club.

But Man must compete and judge his peers, it seems, so a competition it must be.

So … there are all these near-identical jars lined up, each with their numbered certificate of entry. The ambling banter is of modest disclaimers for one's own jars and friendly praise for your neighbour's. To an outsider they all look absolutely identical except for a slight variation in the tone of amber, and the floridity of the label.

The winner triumphed because his colour was superior (don't ask), the clarity was perfect (so were all the others, I would venture), both jars were filled to an identical level, within the stipulated 2mm corner of the shoulder (alright, it may have been 3mm, but quite frankly, who cares?), and, the final triumph, NO honey was found sticking to the soft card pad inside the lid.

What an odd world it is.

* * *

A final point: people usually think of bees as being amazingly precise mathematicians and engineers. After all, don't they make all those millions of perfect hexagons, each with its invariable internal angles of 60°?

Well, no, they don't, actually. They make little cylindrical tubes to lay eggs in, as close to each other as the waxy material will allow. When one row is finished, they align the next row to fit within the dip between any two cylindrical cells of the lower row. I believe mathematicians call it the 'Orange Packing Problem', or somesuch: 'what is the most efficient way of stacking spheres in a given space?' Clearly, you offset the upper row to 'slot' into the lower row.

That's what the bees do, then they stuff up all the intervening space with wax. This, by definition, will produce a block of cells that appear to be hexagonal. Actually, they're displaced cylinders.

Pretty smart, all the same. And as for that Waggle Dance …

# Spuds, Winter 1996

We'd never seen potato blight till we came to Wales, and our first crop of much-loved Desiree was destroyed. We tried spraying with Bordeaux mixture, but it's a soul-destroying job, and virtually impossible to do properly, as you need to spray every inch of leaf, both upper and lower side. What's more, it's expensive, time-consuming, and of questionable 'organicity'. Isn't copper a heavy metal? No doubt it will be banned sooner or later.

Our solution to the problem was firstly to experiment with lots of different 'blight-resistant' varieties, but given the infinite flexibility of the word 'resistant',* we didn't have over-high hopes. None of them appeared to resist The Stinking Death with any great conviction, and all succumbed well before we could lift and store them safely. So the following year we moved on to a more reliable plan, based on a tried and tested principle.

The only fool-proof way we know of solving any problem is to reorganise the component parts and parameters so that the problem doesn't arise in the first place. If the normal or 'correct' way clearly doesn't work, we just stop faffing around and try a completely different approach, and are often pleasantly surprised by what we discover on the way.

So the change of plan was this: instead of sowing $x$ yards of maincrop, guaranteed to be blighted, we would plant $2x$ yards of earlies, which should be up and out and in the bag

---

* 'I think you'll find your wristwatch actually *IS* water-resistant, sir; for up to two and a quarter microseconds, as our exhaustive tests have verified. Did, er, any member of our staff, or the guarantee, ever suggest anything otherwise, sir? I think not. And *you*, sir.'

before the blight could reduce them to sacs of gluey pus. And as earlies take up less room, we might expect comparable returns per foot-row.

We knew Maris Bard was a good all-round performer: it bakes, boils, mashes, chips, and stores well. What more do you want! Oh … and it tastes great too. So we planted lots of Maris Bard, and more or less forgot about the others, 'resistant' or otherwise.

Thus the system evolved. We now muck the land as well as possible and plough, harrow, and rotavate it in (or just rotavate, if appropriate). We then use the ridger to make furrows, sow the seed, and cover by hand (well, 'by rake', actually; no Luddites, we). We earth up either with the ridger, which can be risky in wet weather, as the tractor's bald tyre slithers somewhat and thus tends to skid while the other tyre grips, with the ensuing risk of an agricultural version of looping the loop; or by hand (see above); or by using a little 50cc rotavator we call The Nibbler. We discovered that if you pull The Nibbler backwards, uphill, between the rows, it excavates a shallow trench and chucks soil out to either side. It's tiring and a bit noisy, but quick and effective.

We then do nothing till harvest, except pull thistles and rogue artichokes. Ordinary weeding isn't necessary, as the rows are so close together that the foliage meets over the furrows and cuts out the light that weed seeds need to germinate.

Harvesting is a painstaking operation. Each plant's produce is left in situ to air-dry for a few hours. Then we select from each individual plant-crop either no, or one, or two egg-sized tubers as seed for next year. The principle we work on is that if the plant-crop is heavy, we pick two seed from it; if it is very light, we pick none. That way, we hope to cut out whatever genetic whimsy it might be that causes low cropping. It seems to work, at least partially.

Next, we pick out all the spuds with slug or spear holes,* vole-damage, or heavy scab. They go into sacks marked 'GOOD BARD', for immediate use. Any tubers with any trace of blight, and there are always a few, even on these earlies, go straight into our Blight Discouragement Unit, who crunches them most satisfactorily, and occasionally fragrantly belches her appreciation.

Next, all perfect ones over about an inch across are bagged as 'REALLY GOOD BARD' for storage.

The remaining marbles are chucked into a bucket to be eventually wiped, par-boiled and deep-fried. They are utterly delicious.

Forget McFrazzled McReconstituted McWallpaper McPaste McExtruded McStrips; try quality, is my advice. And don't worry too much about all those extra deep-fried calories. They'll all be gone quickly enough, especially if you invite some friends around. Guaranteed. Then you'll have to wait a whole twelve months for the next batch. Oh suffering, thy name is gardening.

We grow about 700lb of spuds off 450ft/row, including a few Estima and Pink Fir Apple (cooked whole; eaten cold from a bowl, like sweets). They store in paper sacks in the shed. Around Christmas we tip them out and remove any bad ones. Around March we rub the shoots off the last of the

---

* Strictly speaking 'fork tine holes'. Sorry to be so pedantic here, but somebody is bound to write in, otherwise: a long hand-written screed in green ink probably, concerning the place, or otherwise, of the spear in modern gardening, and invariably (mis)quoting the famous gardening passage from *Beowulf.*

'REALLY GOOD' Bard, so they'll keep till the new earlies come in.

Then we plant the seeds. Quality spuds for six, all year round. Easy.

And now, a few words of well-deserved praise for this wonderful plant:

### *The Potato*

*Here's to the spud: the humble potato,*
*That versatile veg that everyone favours*
*That tickles our fancy with infinite flavours*
*But regrettably tends to go straight to the hips.*

*Here's to the tattie: the Andean marvel:*
*The tuber whose music is full-blown delicious*
*That keeps well and cooks well; is very nutritious ...*
*But which puts an unfortunate strain on our zips.*

*You can flip it in pancakes*
*You can dip it in dips*
*You can turn it to vodka*
*And sip it in nips*
*You can boil it and bake it*
*And squish it and mash it*
*Duchesse it and roast it*
*And roest it and grate it*
*And slice it and dry it and fry it as crisps.*
*You can chop it and dice it and cut it in strips*

*It's a wonderful thing, is the common potato*
*And I'm pleased to have written this list of its charms*

*I hope you've enjoyed all the praises and quips:*

*And I'm even more pleased that I've finished this tribute
Without even once ever mentioning ...!*

* * *

The blight took us completely by surprise. We'd read of the devastation it had caused in Ireland in the 1840s but had no idea what it meant in practice. But then to see the healthy leaf of your own staple winter crop developing first one brown spot, then another ... then to watch the rot creeping slowly down the stem, till the whole plant droops over, fatally sick ... and finally, to see the tubers marked with lesions, like melanomas, knowing they could not possibly keep to feed you through the hungry winter months ahead ...

This put us much more directly into contact with what it must have been like 150 years ago in Cork and Kerry, when the failure of your spud crop didn't mean having to spend a few pence extra down at Safeway once a week, but instead the slow and inevitable starvation to death of your entire family and village.*

I think it can be no coincidence that the two worst things a desperate workman can call a strike-breaker are 'scab' and 'blackleg': both cripplingly serious diseases of a staple crop in a society on the constant borderline of starvation. I wonder how or if 'blighter' fits into this pattern?

---

* My family name of 'Griffin' originates from the Cork/Kerry area. My ancestors, perhaps ...?

# Let us spray ...

After the first season we were particularly careful to clear up and burn all the damaged haulm, even taking care to light the fire downwind of the next allotted potato patch. But it made no difference; the blight came back the following year. Ken said he'd never known spuds to have been grown on that field, so it was unlikely to be in the soil itself. Where does it come from? After three years of failing to keep the pest at bay, we could only imagine that it blows in on the Atlantic sou'westerlies that scour and ruffle Pembrokeshire on their way to us. Pembrokeshire is famed for its very early spuds. Q probably ED.

* * *

Spraying was a nasty irritating job. We only had a couple of hundred plants, but we planted the rows close enough together so that the foliage would overlap, to keep down the weeds. This meant that you had to wade through the leaves with your watering can held before you like a particularly ambitious evangelist. The leaves got wet alright, but never wet enough to destroy every cell of blight.

You, the sprayer, also got thoroughly wet from the knees down as you slowly followed the dripping can uphill. Of course it was reassuring to know that neither your calves nor your trousers nor even the insides of your wellies would ever be likely to succumb to blight, but it was cold and wet and chafing, and not as much fun as it sounds.

Things improved when we remembered the big backpack sprayer we'd found at the back of a shed when we arrived. It was satisfyingly coloured in clashing orange and blue, and would carry several gallons of liquor at a time. It had a flexible pipe, and a nice brass wand with an adjustable nozzle on the end. And yes, it worked beautifully, once we'd bought a few brassy and rubbery odds and ends to recondition it. It gave an excellent fine spray, was easily directed, and if the pressure dropped, you just pumped it up again with a little lever.

That little lever could pack in quite a pressure. It idly crossed my mind that if I fitted a couple of drones and little brass chanter, I could have music while I sprayed. But nothing came of my plan, rather to Anne's relief, I think.

This Cooper backpack spray was a great bit of kit. It was much more efficient than the watering can, and made the job sub-tolerable. The only snag with it was that you had to remember not to fill it to the brim, or you couldn't lift it off the ground to get your arms through the shoulder straps. But ergonomically speaking, you really did *need* to fill it as full as you could as it was such a long way to the spud patch and you didn't want to be traipsing back and forth with a pint or two at a time.

So, getting the sprayer onto your back was a two-man job. Or ... as I once surmised, a one-man job with the active assistance of an inclined plane: I filled the tank right up, laid it face down on the coal heap, and then lay back onto it. Arms through straps. Easy. Now try getting up ...

Two-man job, as I say.

# Another fruity bit, but pongier ...

As our Christmas dinner for 1984 began the long slow journey back to dissolution and the sea, we slumped on the sofa and did a rough and ready recap of progress to date.

In 28 months we'd become au fait with the rudiments of livestock management, put up a trio of productive polytunnels, banged up hundreds of yards of fencing, more or less learned how to plough and harrow and rotavate, planted a fair-sized orchard, laid the basis for an irrigation system, learned how to grow a tolerably good harvest from a dozen or so different crops, and found a promising, or at least regular, market for our produce. We were self-sufficient in every imaginable vegetable, from spuds to asparagus peas (which taste like inky blotting-paper, since you ask; and, no, we don't grow them any longer); and in beef and lamb and chicken and eggs; and in milk and butter and cream and cheese; not to mention honey and wild mushrooms (field, parasols and blewits mainly) and hazel nuts. And feathers and dung and wood.

Everything we'd just eaten at the Christmas feast we had raised ourselves, apart from certain elements of Mum's trifle, which was unthinkable without glacé cherries, Bird's custard, and those soggy bread-cake finger things.

We were also full of the joys of achievement and creativity, and a surfeit of healthy exercise and stunning natural beauty, all the way from the wonderful saxe and vanilla skies, to the deeply lustrous and mysterious lapis eyes of a newborn calf.

Were we pleased? Yes.

Were we smug? No.*

---

* Well, perhaps just a bit, all things considered. And it *was* Christmas; and the home-made ginger wine encouraged optimism. Also heartburn, and paresthesia of the tongue, as we later discovered.

How could we be smug? There was so much more to do before we could claim to have established our ideal of a more or less closed organic smallholding system which would feed the six of us and produce enough surplus to pay for our petrol and television license and clothes and electricity. And the other odds and ends we couldn't bodge or make or barter for.

We still needed to sort out what to do with all the excess milk we were getting, and we needed to reduce the cost of the concentrates we were currently buying in for the milking cow. We also needed to optimise our future fuel supply and make sure we didn't become too dependent on any one source. Wood was there aplenty in the cwm, but it was desperately hard work felling and sawing it by hand and hauling it up that treacherous slope to the house. We would either have to buy fuel in, which was very expensive, and not in the spirit of the venture, or do something about improving the wood supply. A chainsaw? A winch?

We also needed to think of some sort of extra income, either by 'diversifying' (on five acres?) or taking washing in, or some better-paying equivalent. Maybe I should learn to play the bass guitar ...?

And, although we enjoyed it most of the time, we really ought to be thinking of ways of reorganising things so we were not both working for 70 or 80 hours a week, every week.

As it happened, Nemesis was looming, but we did not suspect it.

\* \* \*

I nearly forgot: we were more or less self-sufficient in soft fruit as well. In the first autumn, pretty much in the first week we arrived, we had designated the far end of the garden ('The Top Triangle') as a soft fruit area and planned and dug accordingly. I spent much of the September with a big manly spade, hand digging the allotment-sized area, fine-tuning my neglected body in the process. The flab of decades of soft suburban living began to disappear, and strange bumps began to form on my upper arms. We took medical advice, and decided they must be muscles.

It was a fine long Indian summer, as I recall.* Some days it was so warm I worked stripped to the waist; then, when it got even hotter, I sometimes took my shirt off as well. Ah, freedom.

Three other things stick in my memory from this first groundbreaking venture.

Firstly, the endless miles of couch grass that had to be hauled out by hand and burned. I knew how to handle couch, or 'twitch' as it is called in Nottingham. I had once idly complained that our new allotment was riddled with the stuff and that it was hard to get rid of. 'No it isn't,' Grandad boomed. 'You bend down and pick it up.'

That was some of the sagest gardening advice I ever received. Sometimes you just have to put your head down, think pleasant thoughts about the inevitability of Wales (or England) winning the World Cup (again), and plod endlessly on; especially if you're going to be organic and Tumbleweed holds no promise for you, even in the wee small hours of aching bones and post-modernist angst.

---

*An odd term, don't you think? My understanding of an 'Indian summer' is that period in which the broiling tropical sun is hot enough to desiccate and finally crack the eyeballs of any goat that ventures out of its cave. How has it come to mean 'a nice warm autumn'?

The major snag with couch, which I'm sure doesn't need spelling out, but what the hell, is that it's tricky to dispose of. It's as tough as cobbler's twine, and a single inch carelessly dropped will be up and running within a week of the first vernal pulse passing through it. It grows at the rate of about a yard a year. Left unchecked, your carefully cleaned 12 foot bed will be riddled once again in only two seasons.

You can't compost it. At least I can't. I tried once and ended up with a compost heap that looked like a huge ball of pre-knotted string stuffed with coffee grounds. It just *loves* compost heaps.

You can dump it; but where? Only somewhere you have no intention of ever growing anything on. And dumping means wasting all that nutrition within the tubers or stolons, or whatever.

That leaves either dissolving it in aqua regis, which may or may not work, and sounds very expensive anyway, or burning.

You can't burn it green, as it is stuffed with Life itself and is thus 'greener' than the greenest leaf.

That leaves, at least as far as I can see, only one option, which is the one we adopted.

We knocked up a few rough frames of old lath, and stapled some ruined chicken wire across them. Then I hauled each strand of couch out of the earth, like Hemingway landing a particularly difficult marlin, and coiled and dumped it on the frame. The sun did the rest, though it sometimes took days.

When suitably shrivelled and wizened, it would burn. Oh, yes ...

Would it have composted, in this mummified state? I guess it might have, but I've seen too many horror films to even consider it. Safe in bed, quietly asleep. Then the rasping tap at the window …

And then all the shrieking and squealing from the womenfolk and the sending for An Expert who invariably gets delayed, and having to go out and face It myself while children huddle in Mother's arms. Hack hack hack with billhook; tendril round ankle, slowly hauling me into the stygian Unknown; arrival of Expert in nick of time with syringe of experimental serum or a bloody big axe. Either way, safe at last, cup of tea, hugs all round, adoring wife, tearful children, modest shrug of manly shoulders.

Yes, all very well, but it means a lost night's sleep, doesn't it? And I can't be taking that sort of risk. Too much to do.

So we burn it. Crackle, lovely crackle.

I remembered reading somewhere that dried couch root is sold as a folk medicine in the market at Bologna, and vaguely considered investigating further, working on the principle that the best way of getting rid of a problem is to sell it.

The scene at customs could have been amusing. But did I *need* the harassment of being jailed on remand for months while a whole troupe of Experts tried desperately to demonstrate that 32 tons of dried vegetal roots must be a narcotic of some kind? No I did not. So burning it must be.

Oh … crackle, crackle, crackle. Repent! Confess!

The second vivid memory is of an appalling stench that flooded from the earth as my spade cut into another spit. I pulled out the spade, and revealed a shred of black plastic bin bag. We know what it was, don't we? Some previous owner had disposed of his deceased mog with what he thought was suitable ceremony.

Oh, what a stink.

I couldn't understand the thinking behind it. If a critter dies, yes, you bury it, or if of an oriental disposition, dispose of it on a pyre of flaming couch roots, but you do so in the expectation that the tiny corpse will naturally and eventually wish to revert to its constituent elements, in the time honoured manner. It will seek to return to the earth, dust and ashes from which it once mysteriously arose.

But not if it's in a plastic bin liner it won't. It will just stew and casserole in its own juices; and store up an inflated volume of unspeakable pongs just waiting to erupt in the face of the next gardener who happens along.

Perhaps the Egyptians should have considered this approach, to protect against tomb robbery. 'I know, let's just wrap him in a plastic bag! Then, if anyone breaks in, there'll be this huge pong, and … what? Well … what?'

I averted my gaze and nostrils and disposed of the whole voodoo mess into a new hole and hung the bag downwind to dry off a bit before finally binning it. Some things are best binned.

My final memory of the big dig was a decision taken fairly early on to never use a big manly spade again, if a smaller wimpy one would do.

True, the big spade shifted huge quantities at a time, and developed those odd arm-bumps rather well, but it did so at the cost of too many unnatural creaks and popping sounds in the lumbar regions, and I had to stop for a brief lie down every three or four spadefuls. Was this a sensible way to dig a plot?

Why not use a smaller spade, which would shift less stuff at one throw (there's a laugh), but which didn't threaten rupture or spontaneous evisceration.

And, as any student of gearing will tell you, a small cog does the same work as a big one, and needs less energy per throw. Thus, a small spade enables you to do the job perfectly well. It just takes twice as many 'digs', and will, ergo, take twice as long.

Or does it? No! In fact, you can work faster with a small spade because you can build up a helpful working rhythm and don't need to lie down at all. The Way of the Yogi, rather than the Way of the Highway Maintenance Operative.

This was a significant discovery. Now I never use a big tool if I can get away with a small one. This strategy works particularly well with spades, forks and rakes, but not, unfortunately, with spanners or bath plugs.

We planted eight goosegog bushes (as I recall: Whitesmith, Careless, Early Sulphur, London, Leveller, Lancashire Lad, Whinham's Industry, and Gunner); five raspberry types (Malling Exploit, Malling Promise, Zeva, September, and Norfolk Giant); three sorts of blackcurrant (Baldwin, Laxton's Giant and Wellington XXX); and one or two each of redcurrant, whitecurrant, Worcesterberry, loganberry, Japanese Wineberry, and a Himalayan Giant blackberry which disappeared without trace, against all linguistic and horticultural expectation.

We also tried a couple of blueberries (Earliblue and Pemberton) which were probably the most successful plants of them all, as they seemed to appreciate the natural slight acidity (~5.4pH) of our land. Unfortunately, the birds love the berries, and we've never got round to netting them (the bushes, obviously), otherwise we would be picking pounds and pounds year after year.

* * *

So we ended 1984 on a high note, full of confidence and bright shiny optimism.

1985 could only be better, couldn't it!

# House Music, Spring 1997

Our house is a traditional Welsh construction, designed to shut the weather out, rather than to welcome Nature in. After all, if you've been in Nature's rumbustious company all day, you might look forward to a bit of peace and quiet at night.

Thus the doors are small and the windows are tiny. On all but the brightest of days we need electric light to work in comfort, or, indeed, safety; and at a bare 5½ft high, our front door is a low-level menace to the foreheads of visitors. Tonsuring is a constant threat; scalping a possibility.

So we live in what amounts to a cave with walls getting on for three feet thick. Each wall is really a double wall with the central gap in-filled with mud and rubble. The local stone is pretty flaky schist (a sort of half-baked slate) and it would be impossible to build with it in any other manner. Every local building is similar (except the new hacienda-style bungalow-chalets all along the main roads. Wales is just catching on to ribbon development) and they look fine.

Of course, you can't build grand edifices with porticoes and pilasters out of shattering schist. You build a box, then slate the roof, then light the fire, as quickly as possible.

There used to be local custom that anybody who could build a house on common land, and have smoke coming out of the chimney by dawn, could keep the house to live in rent-free. The new householder could then throw an axe to determine how much land he could use for himself. I guess the whole family pitched in with the building, but even so, it would be a marvel of prefabrication, planning and labour to get anything bigger than the barest hovel slung up overnight. No doubt custom allowed for discreet re-jigging and extension when required in later years.

When these farmhouses were first built it would be a rich man who would have a second storey, and only the local aristocracy would ever see a brick. Our house is pretty typical, I think. It looks as though it began as a two-roomed cottage, and a local historian suggested it had been built by the Sheriff of Cardigan in the 1690s. When we stripped the cladding off the back wall, for repair, we could see the stone courses of the original building. The higher courses had clearly been added later, to make two six foot high bedrooms.

The thick walls act as storage heaters, and once warmed up, take ages to cool, so the house is warm in winter and cool in summer: air-conditioning, the Green and appropriate way.

One of the minor joys of farming is to wake up to hear a gale hammering rain down onto the slates a few inches above your head, knowing the place will never blow down. The joy fades when you remember that soon you'll have to go outside to milk the cow or haul a sheep out of a ditch or a tree or wherever its own whimsical nature and the weather have led it, but what the heck …

In springtime, we wake up with anticipation of a day of land preparation or planting, but because our bedroom window is so small that it only usefully registers a binary signal of light/dark, we rely on the birds for weather information. A good lusty bellowing from the blackbirds means a) the day is fine, and b) it is time to get up. No lusty bellowing means c) it's raining (which you can probably hear anyway) or d) snowing (which you can't). So ten more minutes, perhaps …

One particular blackbird sits in the leylandia near the window. For two weeks he drove me increasingly scatty by whistling the first bar and a half of *Diamonds Are a Girl's Best Friend* over and over. Last week he had a change of heart and switched to the first bar of *Rawhide*. Now he seems to have abandoned that as well, because this morning he alternated between the first 1¼ bars of *It's Istanbul, Not Constantinople* and *Putting on the Ritz*. Oddly, this variation was even more irritating.

Why can't he settle on one song, learn it properly, and sing it all the way through? Where is his artistic integrity? And why such safe middle-of-the-road old tunes? Why won't he give something a shade more modern a try? Or something a little more demanding, like *Blue Suede Shoes* or *Living la Vida Loca*?

Or perhaps I should just try teaching him something more appropriate. *This Old House*, maybe?

* * *

I don't think I buy that Sheriff of Cardigan story, actually, unless the house was knocked up as a worker's cottage, or somesuch. We're getting on for 20 miles from Cardigan and I don't see a post-medieval Sheriff commuting 40 miles on a donkey every day. As soon as he got to work it would be time to go home again. He'd be permanently saddle-sore, poor chap, and have zero job satisfaction, on account of having done zero job.

And why would he want to live in a hovel in the sticks, when he must have had the option of a big house in the town with a tafarn next door selling toasted sheep sandwiches, served by strapping wenches with the traditional huge jugs of whatever you fancy?

# Three centuries on …

But the house *is* old, there's no mistaking that. We watched with some wonder as the old cladding came off. Such a little house it must have been once. Just big enough to hide in for the night, and hibernate over winter, with or without the donkey for company.

I wonder who the people were who first built it? Could it have been a young married couple starting off in life together in a 'tŷ unnos', one of those 'one-night houses' that were thrown up in the hours of darkness? No, surely not; small though it was, it looked too big for that. How much land did they have? What did they grow? The geography and climate haven't changed much, so I guess they raised sheep and cattle; perhaps grew oats as well. They'd have leeks and onions, garlic, turnip and cabbage, and some sorts of beans and peas, but nothing like the huge variety we have available today.

Whoever they were, their world would have been unimaginably different from the world of twentieth-century English incomers looking to grow 'organic' veg.

Would 'Thomas and Rhiannon' have had any concept of 'organic', for example? Clearly not. *Everything* was 'organic' in the seventeenth century; there was no alternative to 'organic', so the word would not even have existed. Did they have a kitchen? I wonder if they even had glass for their tiny windows? Unlikely, I'd have thought. Tiles for the floor? Costly things. Who knows?

They would certainly have burned wood, but I imagine the house is just too far from a pit for a small peasant farmer or labourer to afford the haulage on already expensive coal.

Their furniture would be mainly inherited or home-made.

They'd grow most of their own food, of meagre diversity, but probably adequate volume and quality. Their beasts would be hardy but small, as would Thomas himself be. He wouldn't need a six foot doorway.

Would Rhiannon have had a garden? A few local herbs, maybe, to liven up the porridge or cawl, and perhaps more importantly, for medical use. The Physicians of Myddfai used ancient herbal remedies calling for sage and rue, rosemary, marigold, hyssop, primrose, house leek, saffron, broom, thistle, parsley, honeysuckle, chickweed, tansy, and dozens of other plants. Also hedgehog oil, eel's blood, hare's gall, stale urine, and cock brain.*

What other flowers would she have grown? And what would she make of our own drifts of crocosmia and alchemilla, and the blue eucalyptus, and even the dreaded leylandia accursida that the blackbird sings in?

Our world has changed so much in 300 years that it's hard to know what points of contact we might have with the original builders, should we ever meet.

Imagine the faint electric shimmer in the ether as the time-warp delivers the original builders for a day trip to the twenty-first century.

They arrive in their own old sitting room, which they would not recognise at all, except perhaps for the rough oak beams in the ceiling. The walls are covered in textured wood-chip paper, and painted in a satin pastel tint. There's electric light; a finely finished table made of an alien mahogany wood; a fine wool carpet, with complex designs worked into it; a steel

---

* Did they *have* football hooligans then?

stove, with glass windows. The chairs are soft and stuffed, and covered in fine and marvellously finished and decorated fabrics. One chair tilts back if you sit on it. Another tilts back if you press a switch. The sofa is covered in a very odd material that doesn't look entirely natural, very glossy and smooth. And in the corner is a weird shining box with moving coloured pictures in it, and the voices of lots of people who are in the pictures, but all moving too rapidly to make any sense of; and music! Very strange music dashing in and out haphazardly over the voices.

And colour! Colour everywhere! Paintings and drawings on the wall; and colourful and intricate, but quite meaningless, exotic ornaments; wood finished in delicate paints; paper and print everywhere, too, with brilliant colour on every page; books, books; magazines with dazzling coloured pictures and drawings of places and creatures quite unimaginable, and some quite incomprehensible; a finely made guitar. And everywhere you look, dead straight lines and spot-on right-angled corners and surfaces so smooth as to be unimaginably perfect. All very alien to the world of the Welsh countryman 300 years ago.

About the only point of reference we would have in common would be the ash log burning in the stove. And anyway, we'd have very little to say to each other because they would speak no English at all, and our lucky dip of modern Welsh would mean little to them either.

But I expect we'd get by well enough when the dog trotted in and made a fuss of them. Some things don't change.

And things would warm up a treat when we wheeled in the toasted sheep sandwiches and home-made beer; and the peanuts, chocolate, wine, oranges, bananas, tomatoes, crisps, and tea and coffee and the thousand and one other modern commonplaces that we all take so much for granted, but which would be so many marvels and wonders to Thomas and Rhiannon. How unbelievably rich we would seem to them. Far richer than the local bigwigs in their day. Much richer than the Sheriff, even. Princes.

It so happens that as I write these lines I have a glass of rugged cider to hand, sailed and laboriously carted in from the French lands, I believe. And at this moment, I'd like to drink a toast to Thomas and Rhiannon, or whoever they were who built this little house 300 years ago.

What we would really have in common, except for a fondness for this home, is a respect for the land and the creatures in and on it; and thus for each other. Another of the things that never change. Iechyd da!

Was that a bit soppy? Sorry if it was, but it was heartfelt. I wonder if someone will be raising a silent glass to Mr Wimpey or Mr Barratt in 300 years' time? Answers on a rice-paper postcard, please.

\* \* \*

I wonder what an average habitation span might have been back in those days? I would expect the house to have been passed on down the family (unless it was rented out, of course) until there came some very good reason to sell it on. Lack of heirs? Bankruptcy? Realising assets? Emigration? Epidemic?

I would not imagine an incumbent would move out very quickly after moving in. This is the deepest countryside, and populations moved very slowly in the country up till the 1950s. Even now, there are people I've heard of who have never travelled more than 50 miles from where they were born. At least one is apparently proud of the fact.

It's not impossible that there have been fewer than 20 families in this house since it was built. Maybe only a dozen. The parents of each would all just about fit in the sitting room, standing up-close and rather embarrassed, I should think. What a fascinating history they could compile between them.

Who was it who built the big Blue Barn, for example? Was it he who ran the flume down from the pond in the top field that now belongs to the owners of the big new bungalow? That flume would have followed the course of what is now our ever-threatened driveway, and turned a little water-wheel at the bottom of the yard. You can still see the deep slot built into the barn wall for the drive belt off the wheel to go through. What machinery did he have in there? Thrashers? What else?

And who built the dairy, now the kitchen of the farm-house? And the little milking parlour? And the little shed opposite, now April's winter quarters?

The only previous inhabitants we know of were the vendors, who ran cows and sheep on 40 acres; their predecessor, who did likewise; and his predecessor who changed the farm name from something very Welsh to something rather easier to spell for incomers. He was Polish. The deep irony of this is not wasted on me. Have you seen *Polish* spelling?

I happen to know that one of their words for 'beetle' is spelled 'chrzaszcz'.*

All the Poles I've ever known or heard of were eccentrics.

---

* To pronounce 'chrzaszcz', take a deep breath and say 'conch' several times. Then grab a pepper spray in one hand and a chair leg in the other, take another breath, and begin the pronunciation again, but immediately after the initial 'c', spray the pepper firmly up one nostril, while beating the other repeatedly with the chair leg. The resultant word will be perfectly pronounced. Most Poles use this method, incidentally.

I wonder if it was he who thought it was a good idea to run the hot-water pipe round the outside of the house?

Wouldn't it be interesting if we had the custom of providing our houses with their own scrap-book or passport, that would have all the alterations marked and dated on it, and the lines of all pipes and cables and conduits indicated? It wouldn't take long to initiate or update, and would be of great practical use too when you want to drill through a wall, but preferably not through the mains cable or the gas pipe.

Which brings me back to the state of the farmhouse when we moved in.

It was basically sound, and had what estate agents like to call brightly 'lots of potential'. Lots of potential, but only a single, round-pin, five-amp socket in the sitting room, placed half way up the farthest wall from the door, just where common-sense said the sideboard ought to go. Our 12 inch b/w telly flickered bravely from this socket for three years.

The obvious place for the dining room was the 1960s extension that tacked on to the outer wall of the kitchen. The long wall of this room carried the back door and a long window. The west wall had the stove cemented onto it.

The stove was a Tirolia multi-fuel job, from, I think, Austria. It was well made, and other people who know about these things thought well enough of it as a design. Mum, who was brought up on piped gas and electricity in Liverpool, did her best with the brute, but found it pretty baffling to cook on, as a twig here and a briquette there can make all the difference between a soufflé and a paving-slab. Anne, who was brought up in Aga country in Sussex, soon came to some sort of working relationship with it, however.

We also had an electric stove, which meant that somehow or other, we could cook.

Mum and Anne took it in turns; week on, week off.

The main problem with the Tirolia was that it was in entirely the wrong place in the house. First of all, it was on a blatantly outside wall, so half of the considerable heat it generated rushed straight outside. Needless to say, this breezeblock and blutack extension was not insulated at all. Even though the stove was only a few feet from the kitchen, little heat ever wafted its way. We had to rely on the radiator for that, and, frankly, the stove was carrying rather more radiators than it should have if any of them were to operate properly. We could warm up the sitting room by turning off the radiators in the bedrooms, or vice versa. Thus, we had to constantly juggle, and winters were rather chillier affairs than they might have been.

That stove would have to be moved. One day.

What's more, the joke of a chimney poked out of the wall and rose to only a very modest height above the flat roof; which was itself several feet below the height of the main body of the house; which lay between the chimney top and the prevailing wind. Eddies, cross-currents, and whatnot being what they are, that blasted chimney would *not* draw. The fire never got a proper blow across it, and whenever we tried keeping it in overnight, as in the depths of winter, for example, we could expect to come down to a kitchen smeared throughout with a thick blue haze, of the sort that hangs around in Flanders trenches to this day, soaked with sulphur that made us all choke. Gas! Gas! Horrible.

Before you could make a cup of tea you had to open both

doors for ten minutes to flush the poison out. Longer, if it was the sort of still, frosty weather that reduced the feeble draw on the fire to zero. Then we needed to run an electric fan for half an hour to make the kettle approachable. Ridiculous. And *freezing*, don't forget.

On top of this, the end bedroom was still inclined to dampness, despite removing the defunct chimney, and black mould was persistent on the walls of the bathroom and kitchen. There were odd damp spots elsewhere, but nothing worth getting excited about. It's a very old house, and old houses have their eccentricities. And black mould is endemic, we were relieved to learn from our neighbours. You just cope with it. (We did eventually cure it; details another time.)

We'd heard of Improvement Grants and applied for one. A very nice man came round with a clip-board and spectacles, and looked and poked. The deal was that yes we definitely qualified for a Grant, but they would only pay up the Grant money if we had things done their way. Deal? Deal.

Part of the deal was that all the downstairs floors would be replaced with proper waterproof concrete ones. I guess the original stamped earth was still in situ under the cracked and heaving quarry tiles. Also, we must have any damp in the walls treated and inspected, and must have the perfectly sound and harmless asbestos roof of the utility room replaced with slate. And a few other oddments. No problem. The nice man was a real help.

I was muttering to him one day that we had three flat roofs, and that flat roofs were devices of the devil that always leaked, and instead of ripping off a perfectly good asbestos roof, why didn't they help us to replace these awful useless flat ones? He listened politely, then explained to me some detail of the regulations that wouldn't allow them to do this sensible thing.

Then he said, 'But let's have look at them, shall we?' Two were definitely out of his realm of patronage, but his manner changed as we entered the third room, a cold and nasty little modern bathroom, tacked on for no apparent purpose, next to the sitting-room. 'I couldn't help noticing,' he gestured with his pencil, 'that you have a bucket in the middle of the floor. Would that be to catch the water I perceive to be dripping off the light-bulb?' Indeed, it was. 'The ceiling is also bulging quite amusingly, I see,' he continued. 'Perhaps we could incise a small incision? NO ... don't turn on the light.'

A gallon or so later, he was peering with his torch into the hole. 'Good news and bad,' he said. 'The bad news is that the wet has caused a large patch of mould to develop on one of the beams. The good news is that that means I am allowed to approve a new roof which will prevent the spread of said rot into the structure of the house.' Now wasn't that nice?

# Foxes, Summer 1997

Thanks to foxes we're down to our last two ducks. Drakes, actually; so no eggs. In fact they're the very last of our poultry, and they too might well have disappeared by tea-time.

We're all in favour of wildlife, and we do our best to encourage it, but I must admit to no liking for the fox. We must have lost a couple of hundred chickens, ducks and geese to the brutes over the years.

We kept poultry in at night and free-ranged them during the day. That's when they were ambushed and slaughtered; three or four at a time; definitely not killing to eat, but just killing.

More often than not they would be taken within 20 feet of the back door. The first we'd know of it would be a mass cackling. We'd rush out and find a few odd feathers. At evening roll call there'd be two or three faces missing. Perhaps one of them would turn up next morning after spending a terrified night tucked behind a sheet of tin, or huddled in a tree. But sooner or later she'd disappear for good. After one alarum, five hens were killed within seconds in the old slurry pit, not ten yards from where we were working.

So, as far as we're concerned, free range eggs are costing us a pound a piece (feed is expensive), never mind the slaughter of the innocents. It would be cheaper in time and money to feed the foxes direct and cut out the chickens altogether.

We do miss the poultry: the posse of clucking and warbling hens and chicks scratching about the place, guarded by a magnificent cockerel; and the patrol of ducks beating the boundaries, tidying up the slugs. I even (occasionally) miss the mindless aggro from the geese: they have such beautiful, if chilling, blue eyes.

But enough is enough.

I've made one or two half-hearted attempts at shooting a fox, but he's got more time than I have. And it's no fun anyway, sitting motionless for hours in a thicket of nettles, just on the off-chance, getting cramp, damp, and bored rigid.*

So what's the answer?

We do have a local hunt, but I don't like organised cruelty. And they seem to think they have a God-given right to charge across everybody's land, terrifying the sheep and cows, and knocking down fences. Their argument is that they're my allies because they keep down foxes. Well if that's the case, who's killing my poultry in ever-increasing numbers?

A curiosity: three months ago we saw two young foxes at the bottom of our yard. They must have come up from the cwm. Oddly, they were not at all frightened of us, and took some time to leave. Over the following days, Dylan the dog was hyperactive with excitement. He found a large depression in the straw in the cow shed that stank of fox, and later he trapped a young fox in the Black Barn. It was completely terrified, but seemed incapable of planning its escape. Very unfieldwise, for a wild creature. I restrained Dylan, and left the desperate youngster to slink away. It wouldn't have been sporting, somehow ...

We've heard rumours that unscrupulous hunts have been

---

*A friend once told me how his father was being similarly plagued by a fox, so he got himself thoroughly organised and waterproofed and huddled down in some shrubbery to wait. He waited all day. Not a whisper. As the evening dew began to rise, he cranked himself back upright, and took one final stare into the cwm. *Definitely* no fox. So he shouldered his shotgun and turned to leave. Four feet behind him sat the fox, also staring intently into the cwm. It ambled casually off into some bracken before the hunter's jaw had finished dropping.

known to raise fox cubs at the kennels, to be later released for 'sport'. Surely not.

I prefer to believe that our foxy visitors were just taking the first preliminary steps towards becoming fully domesticated, like their doggy cousins.

So, regretfully, the answer to the problem of foxes eating poultry is the answer we so often find effective: Avoid the Problem. Unfortunately, that means no eggs and no feathery pageant. Such is life.

# N n n n n n

So we now had the house more or less habitable, if not exactly primped and papered. The re-rendering helped with the weatherproofing, and the loft insulation helped retain what little of the Tirolia's heat seeped into the body of the house. The stairs were 'made good' by ramming a couple of baulks underneath and nailing. They still creak a bit, but we confidently expect them not to collapse completely.

The new front door was a great improvement too, except that after a few months it began to jam. No matter how ruthlessly I planed it, it jammed again and again. A visiting friend found the fault: the frame fixings had been applied by the book, but unfortunately, not into the actual wall itself, just into the mud infill. Consequently the doorframe was free to move and bow whenever it felt like it. Builders, eh? A fiddling thing to remedy, but now fixed.

The damp course was another point of trial for us. The Grant man rightly insisted that we should have a damp proof course 'where necessary', but didn't see fit to indicate precisely how you fitted a Damp Proof Course membrane into a mud and rubble wall three feet thick.

Really, the only option appeared to be forcing some sort of rough dpc via silicone injection,* and I didn't have high hopes for that, either. How could you possibly impermeably coat or waterproof every inch and grain of mud and slate? And anyway, didn't I hear somewhere that these old walls should not be allowed to dry out completely, or they're likely to fall over?

---

* There ought to be a joke here somewhere involving Pamela Anderson never needing an umbrella, but I'm afraid I can't see it.

Fortunately, the only place that really needed treatment was the utility room wall which was up against an earth bank, and which was of much lighter construction than the house wall.

Unfortunately, I gave the treatment job to the legendary 'bloke I met in a pub'. He turned up in a clapped-out old pick-up with drums of evil fluid and a drill you could find oil with, and pumped gallons of silicone into the utility wall, and all over the floor for no extra fee. We paid up, he gave us a sort of certificate to pass on to the Grant man, who graciously accepted it, and that was that.

Later we discovered that the wall was not 'solid mud and rubble' but a thin rubble wall lined with breezeblock and plaster, and that most of the chemical had gone straight down the gap between the two, as the bloke from the pub well knew, but didn't think to tell us.

Never mind. It seemed to do the trick well enough. Just very expensive for what it was. Or perhaps it was just that the drill holes allowed air to circulate? Who knows?

Later, a neighbour told us about the system he had used, involving a copper wire and some mysterious electrodes. I never took the trouble to follow that one up, probably because I never quite believed him. I wonder what it was?

So the house was now habitable, if rather dim and cold. Further improvements would have to wait.

* * *

Until the fox made the final decision on poultry for us, we were pretty pleased with our duck and chicken regime. Back in Nottingham our Rhode Island Reds had been everything a chicken ought to be: clucky, reddy-browny, frumpy-looking and productive; like a sort of miniature Asian seaside landlady.

But could we do better?

As soon as we arrived we put into practice what we'd read in lots of books and articles and splashed out on some carefully bred Smallholder Specials from a breeder particularly keen on no-nonsense fowls for no-nonsense folks. If you're going to do a job, do it properly; sound investment; the best is always worth the price; all that.

We took the afternoon off to meet the 2.45 from the West Country, and collected our gently tweetling box of day-old chicks. They'd travelled hundreds of miles with no food and apparently no discomfort. Day-old chicks are truly incredible.

They ran round and round like little clockwork dandelions for a while, then sprouted into queer-looking gangly teenagers, all claws and bristles, and finally grew gracefully into proper hens. So far so good. And they produced proper eggs as well. Excellent.

But we didn't notice any particular superiority over our old Rhodies. Perhaps they had advantages that never showed up in our particular circumstances? All we did notice was that they were, as advertised, 'sex-linked by colour'. This was nothing to do with miscegenation or political incorrectness, but rather the fact that male chicks were always white and females always … something else. Black, was it? Pink? Can't remember.

Anyway, this seemed to me to be rather more information than we needed about a hen. After all, it was going to be pretty obvious who was who after a few weeks; ie, at the time when decisions would have to be made regarding each individual's future, or relative lack of it. Some would be laying eggs and going broody now and then; others would be strutting about like ghetto pimps, looking for trouble and stealing cars.

They were the last Posh Birds we ever bought. Instead we learned to go with the flow, and relaxed into serendipity,* and accepted day-old chicks, pregnant eggs, battered old biddies, the halt, the lame, the huddled masses, from whatever quarter they arrived. Some we bought and some were given us. One we found wrapped in a threadbare tartan blanket on the doorstep one morning, with a little pencilled note in a childish scrawl which read 'plese, mister, oh plese look arter me little n what is ill wiv the flux. I must go nah to be wiv muvver …'

Quite untrue, that last one, as you might have guessed. But we did collect all colours, shapes and ages. Sussex, Orpingtons, Red Leicesters, all sorts. They were then free to mix it as they wished.

I once swapped chicken notes with the local blacksmith, a man well into his eighties, expecting wise words from him on the merits of ancient Celtic strains with mystical powers and fine singing voices. Not a bit of it. He went into immediate rapture about his Warrens, a modern breed renowned as hi-tec egg machines. I was quite shocked.

Warrens weren't for us, however. They needed lots of expensive feed and supplements if they were to become true termite queens. Our motley crew would have to earn their keep by scratching out grubs and creepies all day, in exchange

---

* Did you know that 'serendipity' is connected to 'seren', the Welsh word for star? And that 'disaster' is connected to 'astra', the Latin word for star. Curious, eh?

for a small scoop of grain at bedtime, to be fought and clawed over with all the passion of a Harrods' sale.

The only Warrens we actually came across were ex-factory farm specimens that a friend had bought for a pittance as being worn out and useless. They sat apart, tense and twitchy, with clipped beaks, wild eyes, and big de-feathered patches, not knowing what to do with this sudden freedom-stuff which nobody had warned them of. The other hens gave them a wide berth.

The first thing that came to mind for me was Belsen: those bleached-out black and white pictures of the survivors, boney and threadbare, sat apart, detached, confused: worn out and useless.

The arrival of the Smallholder Specials prompted us to fix them up with some decent summer quarters. In the winter, they would be fine spending their nights locked into the old lorry-back by the house, but in summer it would be much nicer for them to have a fox-proof chalet in the field. That way they would be closer to their day's desires, and at night they would manure the spot the chalet stood on. After a day or two, we would move the chalet on a few feet to spread the bounty, and give the chickens a different view first thing in the morning. ('Oh, look, girls! Grass!')

We came up with two designs. One was based on a straightforward adaptation of an existing box; the other was designed and built from scratch. As any builder could have told me, had I had the wit to ask, it is always easier to start from scratch.

The box we got for 50 pence off the local supplier of multi-fuel stoves (the national Tirolia agent as it happens). It was the standard box the stoves were delivered in. I drove down to Newcastle Emlyn in Gloria, our old Transit ambulance, and heaved the box into the back, to the great amusement of the staff, to whom a box was at best just a box, and at worst a nuisance to get rid of.

Back home I sawed out a pop-hole and made up a bit of a door, then coated the lot in creosote. Oh ... that stuff stinks. It seems to burrow into your clothes as well, and coat every fibre with that evocative whiff of Nuits de Kuwait.

With some help from Anne I nailed on a chicken mesh skirt (to the box, obviously), added a mesh floor to keep rats out, and mounted the whole thing on bricks to let air through. Fine. But the hens turned their beaks up and wouldn't go in.

It took some time for me to realise that they also think creosote stinks. So it needed to air for a week or two. Meanwhile the hens bedded down in the lorry back, where they'd left their pillows and sleeping bags anyway, presumably expecting just such a man-up.

The only problem with the Box was that it was just too sturdy. Once the stink had faded to barely tolerable the hens liked it well enough, especially the spacious perching facilities and dinky little ramp to the entrance, but we, the management, couldn't shift the flipping thing around. Too heavy. And inclined to wobble a little round its nether regions, where I had hacked bits out, to remove the 'floor'. This weakness could easily be rectified ... but only by adding reinforcement, which meant making the thing even heavier. Thus the Box was re-allocated as a non-mobile bivouac cum blockhouse, to be placed strategically and permanently somewhere absolutely suitable. Needless to say, we never did find the absolutely suitable spot. But the Box was useful for many years as a sort of Death Row for stroppy cockerels. 'Dead Hen Walking here, so doan gimme no cluckin jive, you dig?'

Plan B was an Ark. Not as in two-by-two, but as in traditional-chicken. The secret was to use modern materials sensibly. It had very slowly struck my attention that a sheet of galvanised iron was just about the right size for the wall of a traditional triangular-section chicken-ark. All it needed was a frame to tack two sheets onto, a back end, a front end with a door ... and there you have it: a perfect summer hen-house. Once the idea had gelled, it took no time to build. AND it didn't need slopping over with stinking creosote. The ridge was easily waterproofed with several strips of ex-wellie, and that was that. Later refinements included extra waterproofing with a row of plastic sacks, improved ventilation, and a nest-box; and a juke-box. No, only kidding. But all the evidence I've come across suggests that you could improve your egg production by 10 per cent or more if you played a little gentle Bach to your chuckies.

\* \* \*

The greatest news of all for the New Year was that my parents decided they'd had enough of being cramped up with the other four of us, all constantly getting under each others' feet, and that it was time to spread out a little. This had always been the original plan back in 1982, though we were not sure what final form it would take.

Building on didn't seem like a viable option, really; at least not on a big enough scale.

They looked at one or two properties in the immediate district, but couldn't find one they liked.

Finally, they applied to build a bungalow in our top field. This required something called Agricultural Planning Permission, and I was very doubtful that we'd be allowed to build, as our total holding was only ten acres, including the driveway and the woodland, and the received wisdom in the Planning world was that you needed at least 40 acres of good pasture to make a bare living. Hence, how could you possibly require an extra dwelling to house a worker on your land if you only had five proper acres?

But then we looked into the matter and saw a ray of hope. What *really* mattered at the end of the day if you wanted Agricultural Planning was not the raw acreage, but the amount of actual *work* the holding represented. Thus, for example, if you had ten acres of roses, which needed close attention all year round, you might easily make a case for there being plenty of hours of work to justify an extra dwelling.

We set to work to make our case. It wasn't at all difficult to do so. Clearly there was enough work to keep three or four people busy all year round. Tunnels, the woodland, the livestock: and in season, an awful lot of labour-intensive handwork on the field. We were also actively expanding our veg into the bottom field. Add to that the potential of a dozen extra bee-hives, and a vineyard, and our plans for more tunnels for radish and salad onions, and the scheme to convert the slurry pit into a fishpond and bring hydro-electricity up from the stream to warm the pond and the tunnels ... We had lots of plans, and they would all require work. Lots of it.

We made a very good case.

The only snags were whether the local councillors in a traditional and conventional sheep area would recognise the labour-intensity of organic vegetable growing, and whether they felt our application could be said to fit in with some plan or approach they were working to.

I was very doubtful on the latter score.

One day half a dozen strangers turned up in the top field and started pacing about. They looked like councillors (sensible clothes; distinctive haircuts). I gathered up some further copies of our Gestetnered supplication and dished them out. They didn't look too pleased.

Oh dear.

Oh dear oh dear.

\* \* \*

Some weeks later I was dropping off a box of veg to a big hotel in Newcastle Emlyn, and chatting with the kitchen staff. The owner came in and smiled and said, 'It went through then.'

Went through? Went through what? What went through? What, indeed, went through what?

'What? Sorry ... what went through?'

'The planning.'

I'll never forget that moment. The blank expression, as I tried to follow him. Then the disbelief. *'Really?'*

'Bet you're pleased!'

Pleased? I nearly wept.

Soon we would all have room to breathe again.

# Shearing, Autumn 1997

Some people will pay the price of a pair of good binoculars to buy a ticket for the ballet, just to watch a chap with very expressive hands and suspiciously bulky tights chucking around what appears to be an anorexic pencil in a petticoat.

Me, I'd rather buy the binoculars so I can sit in my deckchair and watch Ken performing his annual pas de deux with Juliet the Suffolk and the rest of her woolly friends.

Late on a fine June morning, when the dew has gone and the fleece has dried out and 'risen', Ken drives round to our yard and spreads out an ancient and indescribably filthy old carpet. On the edge of the carpet he sets up a stout wooden trestle which has bolted onto it a little Kawasaki engine which locks into the drive end of the clipper unit. This in turn leads via a flexible shaft to the clipper head. A tug on the motor and he's in business. 'Right. Who's my first customer … ? Chas?'

It takes a team of three. One grabs the sheep, one shears, and one packs.

The grabber's job is to convince the sheep that she'll enjoy it really, and then to lead her* onto the carpet and gently upend her for the shearer, then hang onto the struggling beast until the shearer has a firm grip on something appropriate and can set to work.

The shearer and the shearee then perform their complex and entertaining arabesques and tumbles for our delectation and delight.

Finally, the packer stuffs the springy, straggly fleece into a flimsy plastic binliner. A fun job on a windy day.

---

* For 'lead' read: *drag, as in hauling a grand piano single-handedly uphill across a freshly ploughed field*.

Some sheep go gently, some need coaxing, and some refuse; or panic. The grabber doesn't mind panickers too much, as any standing sheep can eventually be moved, but an outright refusal (a sheep on its knees …) is a tricky proposition. A heavy sheep like a Texel can be a real problem and can expect to be called several rude names (many of which may be genealogically quite unfair, it must be said).

The shearer, however, has no love for a panicker, as they can suddenly lash out unexpectedly and violently. I'm afraid Juliet is a dedicated panicker. She'll seem to be co-operating, but at the faintest relaxation of pressure … suddenly elbows everywhere; kicks, scrabbling, coarse grunts; until a knee descends from somewhere and restores the status quo.

It's a joy to watch Ken shear. The client is sat gently on her rump with legs sticking out, like the rear turret of a Lancaster bomber. From behind, Ken trims her neck. Then a swing and a pivot and she's down, having her side and belly shingled (O mind that heaving bosom, barber!). Another pirouette on the heel and he's kneeling across her, shaving great billows of golden nimbus off her back; then another twirl and a twist through 180° and the other side is ready. It's at this point that a panicker thinks she might escape, but just as she raises her head, Ken's instep drops gently to bar her neck. Magic.

When you think that a Leicester Longwool might weigh some 120lb or more, it becomes clear what a skilled operation it is.

It's all a question of balance, of course. It looks simple, and even graceful, because flowing movement is naturally the most efficient. The trick is to keep your own balance perfect, while not allowing the sheep to gain hers. Tai Chi for two.

But give a panicker half a chance …

A moment's lapse in concentration on Ken's part last year, and Juliet turned what should have been an elegant Apache into something more like a guest turn at a punk disco; legs and elbows pumping, violent squirming and rucking, and much vulgar bellowing. A sudden lunge shook Ken's balance, and rather than rick his back again, he released her. I re-grabbed her. Ken started shearing her again. She lashed out again, straight onto the sharp clipper, which nicked her skin. That hurt. You could tell.

But eventually she was shorn and tottered off to meet her mum who was waiting to compare hairdos.

'That looks nice, dear.'

'Bog off.'

Ken's giving up shearing next year. His back has suffered one panicker too many. But his son, Gary, will continue. He's happy to shear three hundred a day. I can't begin to imagine how he does it. His personal record is four hundred in a day. I hope his back holds out. Meanwhile, I'd much rather watch Ken and Juliet than Romeo and Ditto.

# Gated

Shearing is a skill neither of us has ever picked up, because Ken has always been kind enough to come round and do it quickly and efficiently for us. He's a busy man, and it would have been churlish to waste his time while I experimented with mohicans and poodle-cuts on our motley crew, as I surely would have if he'd given me a chance.

We did once do a bit of snippy-snipping, though, with one of those old handshears. It's very slow work, but on the other hand the equipment can't possibly break down or run out of anything, and it's easily stored in the bottom of a drawer and can be carried round in a pocket. An occasional lick with a stone keeps it up and running.

Obviously, you can't snip 300 sheep with it, but a small-holder with half a dozen ewes would probably still find the old shears a useful tool. It also has that added something of the small hand-tool: silence, gentleness, delicacy, and intimacy.

On the downside, you have to hold a panicker down for ten times longer, which isn't good for the nerves of either party.

I once watched a fascinating television programme which showed a collection of Saxon artefacts being lifted from an Anglian bog. Bits of ancient technology like a spear-head and a brooch-pin emerged from the mud, and then a pair of wool shears of identical design to the ones in our kitchen, except that the spring-plate connecting the blades was smaller. How's that for standing the test of time?

And I wonder how the owner came to lose such a valu-able item? One suspects the Vikings had a hand in it.

So 1985 opened with a bright horizon. Work progressed steadily on the new bungalow, and it was going to be finished

sometime in the spring. Everyone was looking forward to having more room.

Meanwhile Daisy went steadfastly ahead with her second pregnancy with us. She'd seen it all before and was not the least bit distracted from her normal daily routine of:

- *stand* up, evacuate bowel, eat, drink, urinate;
- *wake* up, evacuate bowel, eat, drink, urinate;
- gawp out of window;
- chew;
- drink;
- doze or gawp;
- urinate;
- sleep;

… except that she developed a sudden penchant for tinned apricot sandwiches with dill pickles, but I expect that's pretty normal for someone in her condition.

We ploughed and harrowed and rotavated and sowed and planted with vernal vigour. On the new building site JCBs came and went, digging trenches for drains and electric cables and piling up topsoil in a great mound. We persuaded the driver to knock down a section of hedge-topped wall between our field and Ken's so we could replace Dad's home-made SAS-style-stile with a proper farm gate.

The stile worked well enough if you weren't in a hurry; or a sheep. Little hooves just couldn't get the hang of the narrow rungs and kept falling off sideways into the brambles.

This was our first attempt at erecting a gate, and we went at it with equal amounts of great enthusiasm, careful forward planning, and rip-roaring pig ignorance.

To start with we hacked out a nice deep hole with a pick-axe and narrow shovel and then rammed and wedged in a home-grown elm log as a post, setting it as deeply as we could manage, and doing our amateur best to keep it plumb and true. Otherwise the gate would be sure to have a sheep-sized gap under it, which would only encourage wide-eyed limbo dancing, with Cheeky to the fore; or it would swing awkwardly and out of balance, and *grind*, and pull itself or the fixings apart; or it would crash shut behind you before you wanted it to, and whack you on the ankles; or it would bury itself in the mud, two feet short of the closing point; or it would need to be lifted bodily upwards before it would open at all. Hanging a gate isn't as easy as it looks, we discovered quite a long time after we started.

And of course, farm gates exert a terrific leverage at the latch end, so the post has to be rock solid.

Once the post was in, we hauled round an old steel gate we'd found in the bottom yard, which looked to be about the right size, and bought suitable post-furniture to fit it.

Then, after much offering-up and 'upabit no downabit holdit ow ow thatsonmybloodyfootactually' we took a deep breath, sent up a silent prayer, and banged in the hinge spikes, at what seemed to be the right positions and spacing, and as level as a 3lb hammer allowed.

The moment of truth came when we lifted the gate and delicately hupped it onto the spikes.

It fitted!

Better still, it swung!

It didn't jam, or swing uphill, or tip into the mud!

It didn't have an unreasonable gap underneath!

It didn't even skreek or grind!

In a nutshell ... it worked!

A bit of scrap aluminium nailed on the top of the post to keep the worst of the rain off, and there you are: a lovely bit of traditional agricultural technology. Very good. Very proud of ourselves.

All we needed now was something for the gate to latch *into*. Woops.

We bodged something for overnight with a length of baler twine running round a blackthorn in the hedge, and banged a proper fixing post in the next morning. Easy.

The gate has been very useful ever since, for shooing our sheep over to Ken's place when dipping was *de rigueur*, and for Ken to bring his flock through to tack on our land.

# Ruminations

Spring advanced and Daisy's time came closer. We were very keen for her to have a female calf, but she was giving nothing away.

As the poor old dear was getting on a bit when we got her, the plan was that we should first put her to a Jersey bull so she might provide her own successor before we tried any of the fancier crosses we'd heard of, like a Charolais* or Simmenthal.

Her first effort at fulfilling the plan had been Wally. He was a lovely little fellow but suffered from one serious setback: he was male.

Being male is not a good idea in the farmyard; not if you have any plans for longevity, anyway. Never mind being the biggest, finest, roughest, or flashiest calf/cock/kid on the block. If the farmer hasn't got a use for you, you're for the chop/steak/decorative sausage. And in this age of Artificial Insemination, very few farmers have any use at all for very few males.

So our firstborn, the lovely Wally, had to go, at the ripe old age of nine months or so.

This was our first act of premeditated murder on a proper mammal (ie, furry, cute, and LARGE). It was not easy to see him playing and dozing in the sunshine, knowing that sooner or later we'd be sending him to be killed. But we were half-prepared for the decision, as we knew that he was destined to follow the inevitable dictates of his hormones and become huge and ill-tempered and not at all nice.

---

* Not to be confused with a 'Charentais' which is a sort of melon, or a 'Chevrolet' which is a small French milking-goat.

We would have no conceivable use for a modern day Auroxosaurus Rex blundering about the place, ripping up oaks and slinging tractors into the cwm, and we knew nobody else would want him either. But did we have the heart to just 'send him off' one day? Our own sweet little calf?

Wally himself was kind enough to intervene in the decision and seal his own fate.

It is one thing to be bumped and barged by a calf the size of a labrador, but quite something else to be mugged by one the size of an iron-pumping donkey; especially one with two-inch prongs that inevitably catch you just under your ribs (or worse) and unbalance you, and leave tender bruises.

A week or two before he started to actually steam at the nostrils and howl at the moon, Wally went. Ken came round with his trailer and we coaxed him (Wally) with a bucket of oats from the front and a bamboo from behind; then up the ramp, slam the door; and Ken drove slowly up the drive and out of sight.

Then our job was to go inside and try to stay sensible, but feeling awful.

A week or so later we went to collect the meat. Sacks and sacks of prime beef. The best in the world, and just for the cost of the slaughtering and butchering. It's a good thing we weren't hysterical about being vegetarian.

\* \* \*

Then Daisy delivered, as casual as you like on the first day of the new month. Yes … it's a girl! Cigars all round (Daisy chewed hers with some relish) and a bottle or two of rhubarb champagne. For obvious reasons we called her

'April'.* She was the bambiest of calves, and has remained rather more than winsome ever since.

Daisy once again produced floods of rich milk which was too much even for the apparently hollow April to soak up.

Our neighbour to the south had a herd of pedigree Jerseys and sold us a daft-looking little bull calf for a fiver. He was very welcome.

The sad thing is that Jersey bull calves are virtually value-less. Other breeds or crosses are worth the cost of feeding because they fatten up enough to show a profit over feed. Jerseys don't; or not enough, anyway. Consequently, the best this little fellow could otherwise expect from his short life would be to be trundled to Dover in a big truck, crushed up along dozens of his peers, then trucked on to a shed in France where he would be kept in the dark in a box until he was converted into veal, pies, and fertiliser.

With us, at least he had a few months of sunshine and fun, and a sister to play with and bully (le mot juste, I think), and plenty of good grub.

We bucket fed the calves as soon as we could, otherwise between them they'd have cleaned Daisy right out and left none for us, which wasn't the plan at all. We hand-milked, then took as much as we needed for the household, and split the remainder between the calves in two noisy slurpy buckets. The milk sounded even more delicious than it looked, thanks to the brief spells of goggle-eyed bubble-blowing.

* * *

* Cait later told me that if the calf was a male, Anne was planning on call-ing it 'Fool'. I don't remember being consulted about this.

Eventually, the bungalow was finished. Mum and Dad moved up the hill to the freedom and space of three-bed, recep, ktch, and bth; AND a whopping sunroom Dad had tagged onto the south face. Excellent.

We, for our part, brought our own furniture in from its extended holiday in the calf-pens in the Blue Barn, and set about re-organising the kids' bedrooms.

Cait had spent two years in the little boxroom, about 8½ft by 4½ft, and Paddy had been in the other upstairs bedroom, which was of a proper size. Paddy now moved downstairs into Mum and Dad's old room, and Cait shifted across into Paddy's room. Not a moment too soon, as I had almost burned the house down the previous winter by leaving Cait a little night-light on her tiny dresser. Last thing at night I went in to see her, and found the heat of the flame had melted through the polystyrene tile on the ceiling, several feet above.

I shudder even now to think how *embarrassed* I would have been to have burned the house down.

## Hippy Oaf Burns Own House Down. Prat. Pix Pages 3, 4, 5, 6, 7, 8, 9

# We're All Different, Winter 1997

Once in a while someone takes the trouble to contact me, or the editor, to comment on something I've written in this column. On the whole these letters are not overwhelmingly eulogistic. After all, it's a disappointing fact of life that most of us put pen to paper only when we want to complain.

In the past I've heard from people who think I must be mistreating our sheep because of a silly phrase I've used to describe some of our more venerable specimens; 'clapped out old bicycles' comes to mind.*

The most unexpected letter I've had, though, was from someone who thought we weren't fit to keep geese because I called our vicious knife-carrying gander a 'born-again Nazi', or somesuch. The correspondent's argument was that her geese are gentle and wonderful, therefore all geese are gentle and wonderful. This, I'm afraid, is not only bad logic, but it doesn't recognise one of the most interesting of all things about animals: that they are all as individual as people. Every one of them.

This came as a great surprise to me, I must say. We'd kept dogs and cats before, and I knew that they had individual personalities, but I was not expecting this principle to be true of cows, and even less so of sheep.

Our first housecow (Daisy) was bought in as an adult, and had thus learned her primary lessons of Life elsewhere.

She was pleasant enough, but what's the word? 'Wilful', I think. If you said 'right' and she wanted 'left', not even a sack of the proverbial carrots would persuade her. She was impervious to pleading and even commonsense once she'd made her

---

* Or was it 'leaning gently against the hedge like a posse of ancient cardigans in a frost'? No, probably not.

single mind up: 'Look, it's greener this way, Daisy! Look! Greener! Mmmm … Yum yum! See? Nice grass. Mmmm … '

Nothing. Except a sudden lunge to the left, head down. The spirit of the Minotaur translated to a leafy field; with me following said spirit rapidly, on the end of a short chain, desperately holding hat on with other hand, and trying not to let my foot get within range of hers. She's very heavy, and knows it. One determined tread from that great cloven hoof, and I'm off work for a week, sobbing into my Complan while She Who Understands Things has to do all my work as well as her own.

We regularly had to resort to heaving and shoving and shouting, just to ensure that she knew who was in charge.

(It's the same with young children, isn't it? Heave and haul on your Mothercare Choke-Chain: 'Sit!' 'NO! I said "NO!" and I meant "NO"! Get off that settee.' They have to learn who's boss. Law of Life.)

Daisy did become a little more amenable with age, but that didn't stop her from splitting my lip with her horn one day when I (the management) wanted $x$ and she (the Skivvy) was insisting on $y$.

I heaved; she heaved.

I heaved harder; she heaved harder.

I heaved hardest; she instantly changed tack and swung her head back towards me, thus releasing the pressure, and using the strain energy between us against me, whacking me down the cheek from the eye to the jaw with her handlebar horn. Pure judo. Had I not been knocked almost senseless, I might have gone as far as a faint ripple of applause and awarded her an ear. Two ears, perhaps. Preferably her own.

Only the lip was actually damaged; it could have been much worse. I was as much offended as hurt, and I'm afraid

I called her 'a cow'. She leered. You could sense the half-burned Lucky drooping from her lower lip.*

Daisy's daughter is quite different. Placid and docile all the way, and has never shown any Miura tendencies, but unlike her mam she hates being hand-milked. Different, see.

Our neighbour's cows are different again. Bessie likes to lean over the gate to commune with April. They commune about grass, I think. The others hate gossiping with strangers.

The sheep are different according to breed, as well as individually. The two Southdowns, Tubs and Erica, are amiable, homely, stout and chatty. They wear black flowery hats like Grandma Giles, and sensible shoes with a strap and a button.

Bambi and Beauty, the languorous Leicester Longwools, are also friendly, but in a more Isadora Duncan sort of way, and will even follow you around, looking lissom and lovely, and perhaps just a little ethereally distant.

The Southdown/Suffolk crosses are warier. One of them is a real trial at shearing or pedicure time. She splays her legs out as far as possible, like a 1950s occasional table, and will not be moved; her sister feigns co-operation for a long time, then suddenly bolts.

Bella is just normal. Rather too normal, we sometimes think. Perhaps she's repressed.

---

* Our guest of the moment thought I should have the cut seen to, as the edge of my lip looked like a tiny slice of tenderised liver. I thought all I needed was a wimpled damsel to wipe my brow with a scented kerchief, and a nice big shot of nice brandy. No brandy in the house alas, and damsel was having none of it, being totally unwimpled and fully occupied cooking supper. Eventually our guest most kindly drove me to our friend the dentist who most kindly put a couple of stitches in for me. Thanks Rob; thanks Colin. I was on soup, and spitting feathers for several days. Cow.

Delilah, on the other hand, has definitely got a wildcat gene in there somewhere. She very nearly *climbed* over a hurdle once and actually snapped at me when I grabbed her. I slapped her face, the trollop.

Meanwhile Cheeky, the Ancient One, takes it all in and gives nothing away. She's a Jacob, and bright. No teeth left, poor dear, but then she's at least 15. Being an old pure breed she's not had the brains bred out of her by farmers who only want her for her body. She's a born leader of sheep if ever I saw one.

It's only a short step from sheep to realising that even poultry have very different personalities. More on the Cockerel from Hell another time.

Meanwhile, I'm now surprised that I was ever surprised at these distinctions. After all, what is life if not a process of growth and learning and individuation? We all do it; people, dogs and cows and sheep and hens alike. No two the same.

Presumably all rats are different. All snakes. All beetles, slugs and earthworms. All nematodes and greenfly? All protozoa and bacteria? Why not?

Pauli's Exclusion Principle makes it pretty clear that all electrons are pretty individual too.

And I'm afraid Adolf the Gander really *was* a wrong 'un.

\* \* \*

As a society, we seem to be obsessed with the mechanistic view that animals, and even people, are all identical. Could we treat veal calves and battery pigs and chickens the way we do if we were to admit that they are all individual creatures with minds and characteristics of their own? I think not.

Even our own medical system is based on the idea that we are just machines that will respond identically to identical drugs. From practice we know we do not by any means all respond to a drug the same way, but the underlying assumption that 'we are machines and therefore we ought to' is still there. It is this assumption that ultimately prevents Big Medicine taking alternative therapies seriously.

That, and the regrettable fact that an uncertain number of alternative therapists are charlatans, nuts, or both. No doubt we'll get round to regulating them properly one day, and then their qualities will be tested far more rigorously.

# Swine

Pigs: ah, pigs! The only common or garden farmyard beast we were lacking.

People tend to get very fond of pigs in the way others get fond of dolphins. I can't see it myself. Pigs are not impeccably streamlined or graceful or beautiful, and cannot do even a single backward flip in exchange for a small herring (we've tried, so we know).

They also have deplorable manners in the dining department, and stink.

I know it's not their fault that they stink, and that given their head they would wander free in the forest: rootling, acorns, truffles, Esprit de Sanglier, and so forth, dunging freely and unashamed in distant glades and dells; but it's just not practical, is it?* The point I'm trying to make is that the pong is quite suffocatingly offensive, unlike the more homely, and even welcoming, aroma of cows.

And, worst of all, they have pale, sandy-coloured eyelashes, a feature I find difficult in people, never mind pigs. And those eyes ... 'piggy' is the only word for them, isn't it?

Friendly? I don't think so. Greedy? Yes; that's different. Noisy. Unruly. Violently inclined.

It was never far from my mind as I scratched one behind its bristled and flaky ear that the Mafia are said to favour pigs

---

*Speaking of 'Sanglier', have you noticed the huge number of red lorries on the motorways called 'Norbert Detressangle'? The name obviously dates back to the feudal baron 'Norbert de Trois Sangliers', or 'Nobby Three-Pigs' in the modern idiom. Come to think of it, didn't Richard III also have a couple of boars on his heraldic note-paper, tabards, etc? If so, that would make him 'Dickie Three-Pigs', I suppose. Or is it 'Dentressangle'? In which case he's Nobby Three-Pigs-Teeth. Mysteries, mysteries..

as a means of disposing of … Well, shall we say they make the swine an offer they wouldn't dream of refusing?

We drove out on my birthday and bought our first pair of ten-week-old late weaners from a friend, then penned them behind some electric anti-rabbit netting on the field. I can't remember what breed they were exactly, but they were definitely not a rare breed. Quite the opposite; some sort of lean-bred bacon machines, based on a Large White, many generations ago.

Why not a rare breed?

Well, our experiences with the Posh Hens sobered us a little. And we also fell to wondering why some breeds *were* rare. Could it be that newer breeds were *better*?

As our pigs were required mainly for helping out with buttermilk and occasional surplus milk disposal, we were not looking for Good Mothers or Beautiful Natures or Hippopotamoid Dimensions. Just critters that would eat what they were given, weeds and all, and make a half-decent butty at the end of the day. 'Pedigrees make indifferent sandwiches' was the house motto.

When they were 'ready', stuffed to the chops with expensive barley meal, and at what seemed to me to be a scandalously early age, a local enthusiast took them off for us, and shortly returned them, heads and all, in a series of big white (Large White?) cardboard boxes.

Pigs-head sandwich anyone? A nice bowl of warm brawn, then? How about a crisply toasted ear to nibble on? A nostril to sharpen your pencil with?

It was very odd to see these bas-relief profiles, eyes shut, lying facing each other in a box, sandy eye-lashes and all. They looked sun-tanned, or possibly embalmed. Surely not?

In this age of plenty, we didn't fancy crunching our way through a couple of heads, and we couldn't face the prospect of the dog chasing an eyeball across the vinyl either, so we buried the four profiles in the orchard; a double porky Janus. They looked so peaceful.

The meat, meanwhile, was magnificent. We've never tasted pork like it. The stuff in the shops is a travesty. No wonder that as I write, some go-ahead supermarket is talking of selling 'flavoured' bacon: lemon-flavoured, honey-flavoured, and so on. This is the state we've reduced ourselves to via our mechanistic thinking. Next week chocolate-flavoured crisps and kidney-ripple icecream. Possibly even bacon-flavoured bacon, but that might be a step too far.

As for bacon proper ... Anne read it up and got cracking. There are two basic ways of making bacon, it seems; the dry way and the wet way. The wet way seemed to require less attention, so she went with that one:

> To a normal traditionel plasticc coole box, do you add one
> piece or short yard of bacconing meat, with testiccules and
> fragglings scryved out and removed, with also salt and sugar
> brine, mixed according to seventeen pennyweights of brack-
> ish salt per Coventry gill of clean watter, and plum sugar by
> addition and to taste, but no more than a scotch ounce by the
> portion. Let the matter soke until don.

Simple.

I can't begin to describe the quality of the final product.
To try to compare it to the pale wet strips of pink tasteless
flannel you get from the supermarket, that spit and frazzle to
nothing amidst clouds of steam and a thin white scum, is like
trying to compare a Monet to a Stradivarius.*

Anne's bacon was perfection. That's all you can say. You
may never taste its like. We thanked the little pigs in absentia.

We did try a couple of rare breeds once. I forget what they
were, now. 'Small Fat Pinks' I think. They ate more than the
bacon-machines but didn't muscle up as much, so the meat was
more fat than lean. 'Not to modern taste', as we say. Perfect
perhaps for harder times, when bacon fat was the purest form
of scarce energy food available; but not for us. We rendered
down the fat and waterproofed a friend's fleet of barges with it,
and two blocks of flats and a bishop (only kidding).

My final living memory of these two little piggies was of
delivering them to the abattoir ourselves. We borrowed a small
trailer and hitched it to the Volvo. No problem. And drove the
20 miles to the abattoir. No problem.

---

* Stradivarius was a *really* rotten painter.

Then I tried reversing. Ah.

My previous experience of reversing had been with a big muck-spreader behind a small tractor. It doesn't take too long to get the hang of this, and I could soon back with confidence into a gap only an inch or two wider than the spreader.

But a *small* trailer behind a *big* vehicle is something entirely different, isn't it? I wish I'd known this before we set out. I could have practised.

As it was, I provided 15 minutes of harmless if stultifyingly repetitive cabaret for a growing number of fetchers, stunners, choppers, disembowellers, scryvers, butchers and hangers as I repeatedly crabbed my way halfway across the yard, only to lose it when swinging round into the narrow entry to the off-loading point.

Try again: steady, steady … jack-knife! Damn.

Try again: steady … fine … steady … two degrees starboard, Mr Sulu … jack-knife! Damn damn damn grr.

Try again: steady … lovely … check mirror … jack-knife! *%**!*!

A number of God's creatures lived a little longer that afternoon thanks to me, as the entire staff was eventually cheering me on from the top of the steps and exchanging small bets.

Yes, of course we got there eventually, but I had a painful crick in my neck to show for it. And an invitation to repeat my performance at the next eisteddfod, which I declined. A man has his pride.

\* \* \*

The only remaining point of infrastructure was to sort out our hay and straw needs.

Now that our furniture was out of the Blue Barn, we could sell the veal pens (should we have burned them?) so we could use the freed-up floor space for stacking bales onto. Better still, we could split the space in half, stack the bales higher, and keep the cows over winter in the other half. This would give them extra exercise space; would make it quicker to feed and water them, especially if we plumbed in a bath and ballcock which would top itself up automatically; and would make mucking out much quicker, as we could reverse the tractor in and load straight into the spreader, instead of laboriously forking compacted muck out through a narrow doorway.

Great. We had a System. It was pretty ergonomic, more or less self-contained, and just about right.

Onwards!

*My experience of guiding Wally through his first night shocked me into paying a little more attention to the deeper aspects of things than was my normal wont. For the first time in my life, I really began to think. How did he, new-born and dozy, know to butt for his milk? How did Daisy know to eat her own afterbirth, of all things? Instinct? That's just a word, not an explanation; and anyway, where did this 'instinct' come from? And how did Wally come to be at all? Grass in; calf out? An utterly outrageous transformation, by anyone's standards.*

*Then one spring day I was sowing kale and idly prodding at the last remaining seed in my grubby palm, when I was suddenly struck by the astonishing process I was participating in; and the utterly bewildering series of transformations that are set in motion by the sowing of a seed ...*

# The Tale of the Kale, Summer 1998

We've all held a seed in our hand and gazed at it in wonder. How can that tiny hard black sphere convert into a thundering great kale plant with huge crinkly green waterproof leaves, dozens of yards of pale wiry roots, masses of purple shoots, then brilliant yellow flowers… and, ultimately, several thousand carbon copies of itself, each containing the same astonishing powers as its parent? How? Where do all the myriad patterns and designs and energies and potencies come from? No… *seriously*…where *do* they all come from? There's no sign of pale wiry root or yellow polleny flower inside the seed. I've looked. I expect you have too.

Logic and Science require that every effect must have a cause, so all these patterns and forms must have a cause too. They must come from *somewhere*. If not from inside the seed, then they must come from outside. Where else *is* there?

What about DNA? I hear you cry. But as I understand it, DNA is just a template for churning out copies of proteins, not for the shapes and patterns they eventually end up in. In other words, DNA-alone seems to produce tissue (without pattern), not organs (with pattern). And anyway, how can a simple double-spiral of chemicals 'contain' the astonishingly intricate blueprints for the kale's flowering, pollinating, and living-seed-growing systems? How *can* a deaf, dumb and blind chemical contain all that welter of abstract organisation, pattern and information to set up and operate all those interlocking time-based systems of life? To put it rather more provocatively: how can the inert contain the ert?

As there's obviously more to Life than the endless streams of chemical proteins that DNA-alone seems to produce, perhaps we should be asking where all these forces that lie behind those patterns and systems *actually* come from, knowing that we won't be able to answer the question properly, but that we might be able to make a little progress (which must be better than nothing) if we look at the question logically and without prejudice of any sort.

'What is Life and where did it come from' has been driving philosophers potty for millennia. The philosophers have then, in their turn, driven their students even pottier with their exotic and contradictory theories. Thus many people have given up on philosophy as a means of sensible exploration. Just too confusing.

But all is not lost!

There *is* a way to hack through all the confusion … because *all* these endless ruminations can be boiled down to fall neatly into one of only two broad categories. These are the suggestions that:

1. Life somehow arose spontaneously from non-life; ie from inert, non-living Matter/Energy *alone*. This has become known as the <u>Hypothesis of Materialism</u>, because it suggests that 'Matter' came first.

   or

2. Life somehow came first and was somehow involved in forming Matter/Energy. This has become known as the <u>Hypothesis of Idealism</u>, because it suggests that the 'Idea' (ie 'Mind', as part of Life) came first.

(The words 'Materialism' and 'Idealism' (with capitals) are used technically here, and are only loosely connected with the everyday words 'materialism' and 'idealism'. If I had my way I'd replace the technical forms with the much more elegant names 'Stuffism' and 'Mindism'.)

And, in case you've ever wondered, to a scientist a suggestion is called a Hypothesis until it is supported by adequate evidence. It is then called a Theory, and is taken to be 'true' until replaced by a superior Theory which is more inclusive, or 'bigger'.

Now I know that not everyone is immediately grabbed by the idea of checking out the merits of Materialism versus Idealism, and pondering the Nature of Reality and The Big One... so, if you are one of those people, please fast forward to ▶▶, and discover how our new rural environment affected us emotionally, and whether I bought an ice cream or not....

If, on the other hand, you fancy stretching those little grey cells a bit, please read on.... I don't think you will be disappointed...

Another way of contrasting those two hypotheses above is this:

<u>Materialism</u> says that Life, Mind and Consciousness came from Matter/Energy ALONE.

<u>Idealism</u> says that Matter/Energy ALONE is inadequate to explain the existence of Life, Mind and Consciousness, and myriad other worldly phenomena, like 'memory', 'love', 'intelligence', or 'purpose' for example. An additional factor of some sort is required to explain these things.

The two Hypotheses of Materialism and Idealism are quite incompatible. If one is right the other must be wrong. But which is which? And how might logic help us to decide? It seems to me that at every point, logic vindicates one of these Hypotheses and knocks the stuffing clean out of the other one. I wonder if you'll agree?

I see the issue as being like the two pans of a set of scales, teetering on a diamond fulcrum. I invite you to weigh the evidence I offer below, and decide for yourself whether each piece should be placed in a pan. Just one single piece will tip the scales to an absolute decision. Big stuff....

# A little thread of logic

Before Darwin, various specialised forms of Idealism held sway, propagated by the Churches. Since Darwin, the Sciences have propagated Materialism. Where once the Big Idea ruled, now Matter matters.

A hundred and fifty years on from Darwin, modern society turns to a Materialist–Scientist to explain a strange light in the sky, not an Idealist–Priest.

But while I'm mighty troubled by the incomprehensible ramblings of priests, I find I'm mightily troubled by the bland pontifications of Materialism too.

*For example:* Materialism requires the spontaneous generation of Life from non-life… which Louis Pasteur, a hero of the Scientific Method, famously proved doesn't happen. Straightaway we have a paradox: on the one hand 'yes'; on the other hand 'no'. *And 'paradox' equals 'non-sense'.*

Likewise, Materialism requires Mind to arise spontaneously from non-mind, and Consciousness to arise spontaneously from non-consciousness. Something coming from nothing, again and again. Paradox upon paradox. *Non-sense upon non-sense.*

*What is more:* in order for the simplest replicating organism to spontaneously 'arise or emerge' by complete accident from totally inert chemicals, the Hypothesis of Materialism requires an astronomical number of atoms to have accidentally come together in a phenomenally complex and precise sequence of patterns and orders: a process so statistically unlikely, as to require odds of unimaginable squillions to one against.*

These odds suggest that, even if possible in theory, the spontaneous creation of Life by accident is overwhelmingly unlikely. In fact the Cambridge astronomer Sir Fred Hoyle compared the likelihood of Life having ever spontaneously arisen via the accidental assemblage of all the millions of atoms and molecules required… as being similar to having a tornado sweep through a scrapyard, and leave behind it a perfectly formed Jumbo Jet.

We are talking *stupendously* unlikely here. Winning the Lottery is a cast-iron dead cert by comparison.**

Yet scientists who regard playing the Lottery as a mug's game continue to implicitly accept these ludicrous odds by insisting that Life definitely arose by spontaneous chance, upon chance, upon chance, upon chance-to-the-power-of-'*n*'… and by chance *alone*.

*Also:* there is a fundamental law of Science which states that Energy can be neither created nor destroyed. Thus it follows that

---

* For the benefit of any budding bookies out there, who know that odds of 10:1 against are never worth taking, a mere trillion to one against looks like this: 1,000,000,000,000 : 1. Any takers? Now try this: Professor Paul Davies suggests odds of 10-followed-by-forty-thousand-zeroes : 1 against just the necessary proteins forming by accident, never mind the rest of the stuff. By comparison to these 40,000 zeroes, by the way, a trillion contains just 12. As a rough guide, 40,000 zeroes would take up over 20 pages of this book.

** Not *absolutely* true, obviously: but, as the budding bookies will have already calculated, the odds on you winning the Lottery would be something like a hundred thousand to one *in your favour*, (based on a mere trillion to one, never mind 10-followed-by-forty-thousand-zeroes to one) by comparison with the odds against the spontaneous formation of Life from non-life. What? Still no takers?

Life, *if* it is an Energy, can be neither created nor destroyed. Thus Life, *if* it is an Energy, must thus be eternal.

Let this be very clear: *if* Life is an Energy, then by the terms of the Law of Conservation of Energy, a cornerstone of Science itself, Life must be eternal; yet no Materialist will accept this logical necessity (see box below for more details).

---

As Energy and Matter are taken to be different phases of the same thing, according to Einstein's formula $E=Mc^2$, it is taken as fact that E and M must be interconvertible. Thus *if* Life is an Energy, it must not only be eternal (as above), but it might (or must) be interchangeable with gravity or electricity on the Energy side, or a chunk of rock or a beer bottle on the Matter side.

I have never heard anybody, scientist or otherwise, claim that Life is interchangeable with normal physical Energy or Matter in this way. Thus, it seems we already, if tacitly, regard Life as being in some way unique. We certainly agree on it being of a higher order than un-living, lumpen, 'stuff'.

If it is indeed the case that Life is unique, and of a higher order than lumpen stuff, then Life may not be treated according to the rules for everyday Energies and Matter.

And we may *not* reasonably claim that Life, a higher-order and eternal 'Energy', arose spontaneously from 'normal' un-living, lower-order Matter/Energies. Yet Materialism *does* claim just that: that the Higher spontaneously arose from the Lower.

---

Alternative to all the above, if Life is *not* an Energy, then it is something automatically outside the purview of Science (if we understand Science to be 'the study of Matter/Energy and the physical world') and Science can thus make no sensible or valid comment on what Life is, or where it came from.

We thus arrive at more or less the same conclusion, whether we take Life to be an Energy or not an Energy: Life is *different*; it cannot be weighed or measured or inter-changed in any way 'expected' by Materialist Science. Therefore Materialist Science may make no valid declamations as to the nature or origin of Life. (Idealist Science may have a great deal to say, however, and should do so as loudly and as soon as possible, I suggest.)

As a matter of interest, Charles Darwin, the hero of the arch-Materialist 'neo-darwinist' school of thought, made it quite clear in the final edition of *The Origin of Species* that *he had no views* on the origin of Life itself.*

It would thus be true to say that Darwin would not be a neo-darwinist himself.

*Furthermore:* a scientific hypothesis needs evidence to support it, otherwise it is (or should be) discarded.

I have read a lot of books and asked a lot of scientist friends and have been unable to find any evidence at all that Life spontaneously arose from chemicals and Energy (the only sound proof for the would-be Theory of Materialism), and, it seems, *this evidence does not actually exist.* Lots of hopes and assumptions and expectations exist: so do lots of exhortations and reassurances and appeals-to-commonsense by big-shot scientists; but no actual evidence. *None.*

Oparin, Urey, Miller, Kornberg, and all the thousands of other dedicated and ingenious researchers who have been

---

* *'I may here premise, that I have nothing to do with the origin of the mental powers, any more than I have with that of life itself.'* Charles Darwin: second sentence, Chapter 8, sixth (final and therefore definitive) edition of The Origin of Species. Does this quotation surprise you at all? If it does, I wonder why? Might you have been previously misled? And just for the record, Darwin mentions 'the Creator' (yes, with a capital), nine times in the final edition of 'Origins'. Check it yourself. But make sure you are reading the '6th, final' edition. There has been a recent flurry of reprints of the first edition, the only one which does *not* mention 'the Creator'. Now why would anyone want to reprint the first and not the last edition?

trying for the last 80 or so years, have produced a number of interesting and apparently relevant molecules from various pre-biotic 'primordial soups'... but *not* Life; and nothing remotely resembling Life.

In fact, in the 1970s, Oparin, who got the primordial soup ball(!) rolling, suggested that we should give up even looking as we'd found nothing in 50 years.

Whereas physicists tend to dump unproved physical theories within a year or two, Biology has hung on to the entirely unsubstantiated Hypothesis of Materialism for over 150 years, and shows no signs of changing its collective mind, despite the heroic efforts of Rupert Sheldrake *et al.*

The other Sciences, taken broadly, go along with this same Materialist stance as a default. It is never impartially or publicly questioned.

Say 'Scientist', say 'Materialist'.

*What we have, then, is a scientific system, acting as guru to the whole Western world, which bases itself on a Hypothesis which is completely paradoxical, and therefore irrational and fatally flawed, and which is supported by absolutely no evidence.*

Incredible, but true.

Thus, our society, which now takes its cues from Science, in the same way it used to take them from Religion, behaves as if the Hypothesis of Materialism were a proven fact when it claims that Life is an accident, as are Mind and Consciousness and that there can thus be no purpose to anything in the universe (including your life, dear reader), as everything in the universe is accidental and thus pointless.*

---

* By this same logic, of course, these scientists' own jobs and opinions must all be pointless too. I wonder if they've thought of that?

It is thus not surprising that we are floundering in a sea of pointless materialism (with a small 'm'), mistaking gratification for happiness, and possession for purpose; and becoming more and more suicidal and drug-addicted in the process.

'Shop till you drop' sums it up rather more neatly than the coiner of the phrase intended.

And gross pollution and ecological abuse are natural extensions of the whole mad circus.

And all this because scientists who, of all people, should know better, have insisted on defending the Hypothesis of Materialism with all guns blazing, despite it being philosophically irrational, and breaking all the deepest rules of scientific methodology.

It's amazing what can grow from a grain of kale.

---

Please note that all that the arguments in the previous pages do is to suggest that the Idealist 'extra factor' must exist.

The arguments say *nothing whatsoever* about the nature of the 'extra factor', except that it is *not* Matter/Energy.

There is also, however, the unavoidable logical implication that if the 'patterns and designs and energies and potencies' mentioned in the opening paragraph do not reside in Matter/Energy, then they *must* somehow reside in that mysterious 'extra factor' the Idealist says must exist.

Please note also that the arguments do *not* have anything to say regarding the process of evolution or the importance of Science, nor do they comment on such curiosities as virgin births, big bangs, transfigurations, quantum mechanics, or men in purple frocks.

# ... and a little thread more ...

All that the tirade in the previous pages arrives at is questions, not answers. The fundamental question arising is: What *is* this 'extra factor', ... and who will investigate it scientifically if Materialist science refuses to?

It seems to me to be very clear that Science should and eventually must study this issue; but before it *logically* can, by the lights of its own fundamental philosophy, it must undergo a paradigm shift of Krakatoan proportions, away from Materialism and towards something entirely more rational: a simple basic Idealism, untainted by religious dogmas of any sort.

There are no great signs of this vital paradigm shift yet, but it will happen one day; there are already rumblings if you listen carefully. Truth will out, as they say, even in the ivory towers.

\* \* \*

I know well that the above diatribe and comments will not go down well with some readers. One of the 11 responses I got to the original magazine article was a scalding letter from someone who thought he was being a True Scientist by assaulting me tooth and nail for daring to challenge any aspect of the Hypothesis of Materialism. We exchanged several letters, very time-consuming on my part, in which I put the rational objections as clearly and quietly as I could, but to no avail. The attacks just became more and more rabid, with no attempt at all to address or answer the points I raised. He'd completely lost sight of the essential principle that Science is a *method*, not a dogma.

\* \* \*

If you fancy investigating this mysterious 'extra factor' yourself, might I offer you my own terms of engagement to consider?

1. Reason and logic must apply at all times.
2. Any 'law', if applied at all, must be applied universally.
3. No paradox is ever acceptable as 'explanation'.
4. Every assumption must be tested for evidence and internal logic.
5. No dogma of any kind is acceptable.
6. All 'evidence' is to be tested; none is to be rejected a priori.

I originally called these guiding rules my 'Six Principles of Investigative Thought', which sounded rather over-grand, so I abbreviated them to the rather more romantic 'SPIT'. It seems to me that SPIT is all you need for clear thinking.

You might start by asking 'if Life, Mind and Consciousness are *not* physical Matter/Energy… what *are* they? What is the nature of that mysterious and potent "extra factor"… what may be known about it?'

May we assume that a Shaman, a Spiritualist and a Shi'ite will each insist that they exclusively know the answer to this? And a Druid and a Druse? And any member of any of the hundreds of Christian sects? And what about Yogis? Buddhists? Taoists? Voodoo fans? Clearly they can't all be right; and maybe they are all wrong. Or they may all be *partly* right…

And what about the findings of 100 years of Psychical Research? Is it just a coincidence that such illustrious scientists as Sir William Crookes, Sir Oliver Lodge, Alfred Russel

Wallace (the co-discoverer of the theory of evolution), Carl Jung and Sigmund Freud were all interested in it?

Whatever, they all seem to be at least addressing the issue in some way, rather than ignoring it or denying the possibility of its existence, which is all standard 'science' does.

A little summarizing puzzle to end on:

* * *

▶▶ If you've fast forwarded to this point, you'll be glad to know that all you missed was a long and elaborate proof that Elvis was definitely a spaceman, who almost certainly built the Pyramids, Stonehenge, etc., thus teaching Man the secrets of agriculture, writing, pottery, and so forth, by means of a secret language, now encoded in the scratches on the Plain of Nazca. So that's that. Now let's get on: let us retourne, as the French would say, to nos moutons. Or, to be more precise, nos cochons.

By the summer of 1985 we were settled down rather well, pigs and all. My parents ('Korki and Poppet' to the kids) were happy with their new house and all the extra space to call their own, and we were gradually getting the farmhouse sorted out to our own liking.

Paddy had just finished his first year at what was to be his last school, and Cait was a rumbustious six-year-old, happily ensconced at the local primary school. She spoke fluent Wenglish by now.

For our part, Anne and I were beginning to feel like we really lived here, and had got to know a few neighbours. Everyone was very helpful and friendly, lending us gear and helping to get the tractor going and so forth, and offering all kinds of vital advice.

We were also just about beginning to see the world in a different light.

The first time it really struck me that I was now living a different life altogether, *and that it was OK,* was the day I collected the Tirolia box to turn into a hen-bunker.

After I'd loaded the box into Gloria, I strolled into downtown Newcastle Emlyn to buy some shoe laces (baler twine frays rather badly after a while, and if you cauterise it, the blob won't go through the eyelet, so eventually you are forced willy-nilly back into the seedy world of commerce) and was struck yet again by an odd feeling of guilt. Here I was, at 10.30 in the morning, wandering freely about ... and it felt decidedly peculiar; like a prisoner on parole might feel. Surely there was a beadle about somewhere, a nark, ready to pounce and lead me off?

All my life I'd been working to someone else's timetable. For 39 years I'd been diligently present at a time and place decreed by an external authority. I did a mental check ... 10.30 am, Tuesday ... that meant I should be in a class room in the tower block, spouting fraudulent theories of communication, or worse, teaching the mysteries of the semi-colon to the semi-conscious. And now ... now I was free to buy an ice-cream, or make full use of the zebra crossing (Newcastle Emlyn's major light show attraction) several times in a row, or just stand and stare.*

---

* Of the three options I chose an ice-cream. But I hurried it.

It took a whole year for that guilty feeling to wear off completely.

We also began to slow down mentally, in the sense that we chose not to rush around so much. In the city and suburbs you belt from A to B to C 'doing things'. If you can belt to D and E as well, and thus do more things, that counts as success. Tired and happy at the end of a busy day of doing things.

What things? Worthwhile things? Possibly. Er ... not always, actually.

Why the rush? Er ... dunno. Lots to do! Er ... sort of thing.

Would it matter if you slowed down a little and enjoyed the journey rather more, and paused to look at the roofline, or the sky, or the shimmer of the leaves? Er ... dunno, really. Who'd do all the things?

Now we had a brilliant sky any time we wanted it (unless raining; in which case, 'active sky'); greenery; trees; birds; beasts; running stream; woodland. What on earth is the point of rushing? If a job has to be done, you do it. But you don't need to shave 20 seconds off the time it takes. You learn to actually enjoy the job, whatever it is, partly as a deliberate act of emotional hygiene, but also because you can see the point of doing it. Even clearing a drain in the rain can be fun, if you go at it with the right attitude and clear sense of purpose. Live in the moment; carpe diem; carpe secundum.

Slowly ... we unwound. We found the time to stand and stare, and it was good.

Meanwhile, we were also learning the arts of independence. We had both gone straight from school to university, then straight back out into teaching. We were upmarket battery persons, unused to freedom and the possibility of running our own lives just as *we* wished: rationally.

Now we could start the day's work when it suited us, if you didn't count having to wake to the alarm, so you could haul the kids up and out and off to school.

What shall we do today? Well, let's see ... is it raining? Yes. OK, we'll tidy that barn up or saw firewood, or muck out the polytunnel, or paint the ceiling of the kitchen. Or jump up and down in puddles for a bit.

Stopped raining? What's urgent? If it's something vital like ploughing, that takes priority. Otherwise, we can plant out, cut brambles back, yank ragwort, do a bit of fencing, paint a gate.

Best of all, we can pick and mix to quite a degree, so we don't get bored silly with handweeding carrots for eight hours on the trot, or get stupidly exhausted whacking posts in.

It was just great. And we could work together, which we'd always enjoyed doing.

Any clouds on the horizon? the soap-addict asks. Sorry ... no. Not yet.

After three years on the land, our plans were working out rather well, and our efficiency and income were rising every year. In three or four more years, we would have reached and probably passed our 'sensible target' for income: about 40–45 per cent of what we needed in suburbia, with a mortgage from Albatross Financial Services around our necks.

We were feeling emotionally and physically healthy, and were using our minds more creatively than ever before.

We no longer felt unbalanced; we were matching mind to body, and felt fulfilled and strong and healthy. We were getting to cope with the traumas of animals getting ill, or dying on us. We saw Life and Death as more of an integrity than we had before.

'Integrity' was a good word: we had an integrated life as well as an integrated lifestyle, to use a word I normally loathe.

And those arm-bumps were getting bigger and sleeker.

# Nemesis parks her bike, and ...

As Christmas approached, we were feeling decidedly optimistic. A new phase was beginning.

We had our full complement of animals and equipment; the kids were settled and happy; my parents had their own place, which gave them more room; we had more space ourselves; we'd applied for membership of the newly founded organic marketing co-operative, and for the necessary Soil Association symbol, that would enable us to sell our produce as legitimate organic gear. Nineteen eighty-six was going to be just *terrific.**

During the second week of December, I developed a bit of a cold. It wasn't like any other cold I'd ever had. My nose actually dripped clear fluid, cartoon-style. Clearly a bit of nasal soldering was required, then that would be that.

I carried on working, of course, for four reasons:

a) Well, you do, don't you? Flipping cold? Pah!

b) The animals still needed attention whether we were discomfited or not. Milking and feeding the cow twice a day, and counting the sheep before bed time; and the teeming hordes of poultry needed waking up to slop out in the morning, then rounding up and feeding, and banging up again at night.

c) Christmas was almost upon us. Then I would take a couple of days off from anything other than essential work. If necessary, Anne could do the essentials on her own for a couple of days.

---

* You can actually feel Dour Nemesis brooding and pacing slowly up and down just over the brow of the hill, can't you?
'All very quiet tonight, sir.'
'Too damned quiet for my liking, Sergeant ...'
Any moment now ...

d) Paddy's new bedroom needed doing. The pine/gable wall still tended towards damp, so it required battening and tar-papering or somesuch. It would be nice to have it done for him for Christmas.

A few days after the dripping nose business began I was packing Swiss Chard in the packhouse, ready for delivery to a wholesaler friend who took as much as we could offer off to market in Swansea or somewhere further east. Packing chard is not the most demanding job in the world. You pick it up, check there's no slugs on it, weigh a few leaves on the scales, and shove them into a polythene sleeve.

But I could feel it beginning to be too much. Anne came by, and I remember saying 'I can't do this' and made my way slowly indoors.

I went to bed and stayed there. Ten days later Christmas came, and I managed to get as far as the chair in the sitting room, but could not summon the strength to walk the hundred yards or so uphill to Mum's Christmas dinner. So the meal was ferried downhill on trays, though I can't say I remember anything about it.

All through Christmas and beyond, I too was ... 'beyond'. Totally exhausted. Anne says I hit the bottle at one point, and was more than usually sorry for myself, but I don't remember that either.*

What a difference from the previous Christmas! But it couldn't last for ever, could it? Sooner or later I'd bounce back. Illness is just Nature's way of saying 'slow down', isn't it? (And,

---

* Well, you wouldn't, would you? Ed.

by extension, 'death' is presumably Nature's way of saying 'stop'. Hmmm.)

But I remained exhausted right into the New Year, sleeping all hours, drenched in night sweats, and smelling rather worryingly of vinegar.

As soon as the doctor opened up the surgery again, Anne drove me round, then hauled and prodded me into the appropriate rooms and seats. I knew now how Daisy and Wally felt.

I'd never seen a doctor look baffled and very concerned before. She asked questions, but I couldn't string one word together, so she took an armful of blood for tests and told Anne to tip me back into bed.

They tested me for leptospirosis, which farmers can catch all too easily off rat pee in the straw bales, and brucellosis, and diabetes and thyroid problems, and heaven knows what else. All negative.

I was ferried down to the local hospital for a liver scan, and an examination by a skinny little consultant all dapper in a morning suit, would you believe, who looked and spoke alarmingly like Kenneth Williams. I can clearly remember him warning me against the evil of drink, and did I realise that beer contained 10 per cent alcohol? I was too gormless to tell him he was talking 26 carat tripe.

Still no diagnosis. The doctor said 'Well we've tested you for everything bar downy mildew and the plague, and we're none the wiser. It must be M.E.'

M what? I'd never heard of it. I later learned that it stood for Myalgic Encephalomyelitis, which means in plain English 'Something Adrift in the Brain and Muscle Department'. Not a very helpful diagnosis, I thought, especially as there is no treatment, and a sizeable chunk of the medical world doesn't even believe it exists.

To add insult to injury, I also discovered that this condition was currently known as 'Yuppie Flu'. Oh, the humiliation.

So … back to bed, with an '-itis' that took more energy to say than I had available to say it with.

And there I stayed.

Meanwhile, to add to the froth and frivolity that is Yuletide, both Paddy and Cait developed dreadful head-clagging colds, which left Anne as the only one standing. They were both off school for a fortnight. And then, by way of icing on the cake, Porky went down with something doggy and had to be extensively and expensively serviced by the vet.

So we entered 1986 firing on only one cylinder, while the other cylinder tried to figure out the mysteries of snooker by watching endless tournaments from his bed on a very grainy and tiny black and white telly. 'For those of you watching in grey and grey, the light grey is behind the slightly lighter grey.'

Were we downhearted? No!

Not even a little bit? Well ... perhaps just a little bit. Anne had to do all the spring rotavating, which wasn't good for her wrists (or the rotavator, as it turned out), and all the sowing and planting, and everything else the plants needed. And she had all the housework, and the kids to look after as well; not to mention the beasts, the fowls, the cats, the dog, and me. I can't imagine how she did it all, but she did. I am still lost in admiration.

So, for me at least, spring would be a little late, this year ...

By May, I'd begun to recover, and by June I was hurling myself back into work, making up for lost time hand over fist ...

## Hippy Gets Out of Bed: Exclusive Turner Prize Committee Notified

We worked like demons and the crops poured in. Efficiency was up, and our income was set to rise accordingly. Excellent!

Things could only get better, now; couldn't they?

Sorry! I clean forgot to mention the
Purple-Sequinned Sardonic Warbler and
the frightful mess in the bathroom after all.
 Never mind.
Next time, perhaps.